The Measurement of Grain Boundary Geometry

ELECTRON MICROSCOPY IN MATERIALS SCIENCE SERIES

Series Editors

B Cantor and M J Goringe

Other books in the series

Electron Microscopy of Interfaces in Metals and Alloys
C T Forwood and L M Clarebrough

Forthcoming books in the series

Atlas of Neutron Backscattering Kikuchi Diffraction Patterns
K Z Baba-Kishi, D J Dinglay and V Randle

Electron Microscopy of Interfaces in Epitaxial Layers and Artificially Layered Materials
A Petford-Long and J Hutchinson

Electron Microscopy of Quasicrystals
K Chattopadhyay and S Ranganathan

Electron Crystallography of Organic Compounds
J R Frier

ELECTRON MICROSCOPY IN MATERIALS SCIENCE SERIES

The Measurement of Grain Boundary Geometry

V Randle

University College, Swansea, UK

Institute of Physics Publishing
Bristol and Philadelphia

British Library Cataloguing-in-Publication Data

A catalogue record for this book is available from the British Library.

ISBN 0-7503-0235-6

Library of Congress Cataloging-in-Publication Data are available

Published by Institute of Physics Publishing, wholly owned by The Institute of Physics, London.

Institute of Physics Publishing, Techno House, Redcliffe Way, Bristol BS1 6NX, UK

US Editorial Office: Institute of Physics Publishing, The Public Ledger Building, Suite 1035, Independence Square, Philadelphia, PA 19106, USA

Printed in Great Britain at Bookcraft Bath Ltd.

Contents

Foreword

Materials science has evolved as a crucial engineering discipline during the last 20 years. The basic approach of the materials scientist is to investigate the microstructure of a material, so as to optimise its manufacturing process and resulting properties for subsequent engineering use. Electron microscopy has proved to be by far the most powerful technique for examining and understanding material microstructures, and electron microscope methods are essential for developing new engineering materials of all types. The objective of this series of monographs is to provide overviews of the impact of electron microscopy in different branches of materials science. The series is designed to be broadly based across the materials spectrum, including metals, ceramics, polymers and semiconductors, and to deal with the full range of available electron microscope techniques, such as electron diffraction, lattice imaging, scanning electron microscopy and microprobe analysis. In general, individual monographs will concentrate on a particular type of material or a particular problem in materials science, and will review the use of electron microscope techniques to characterize and understand the relevant material microstructures. The series is intended to be of interest to a wide variety of academic and industrial research scientists and engineers.

In this book, Dr Randle has achieved these aims in an authoritative and comprehensive manner, covering all aspects of the use of electron back scattering in the scanning electron microscope in the study of grain boundaries and interphase interfaces in metals and alloys. We hope that the book will act as a handbook for practicing electron microscopists studying grain boundaries and interphase interfaces, and will help all who read it to appreciate the power of the scanning electron microscope in that field. The book is complementary to another in the series, by Forwood and Clarebrough, on the use of the transmission electron microscope in similar studies.

B Cantor and M J Goringe
Series Editors

Glossary

Angle/axis of misorientation θ/UVW, which may alternatively be given in spherical coordinates, p_1, p_2, p_3. Description of a misorientation as a rotation through an angle θ about an axis UVW which has the same indices in both neighbouring grains.

Angle/axis pair, see **Angle/axis of misorientation.**

Asymmetric domain of Euler space. A subvolume of Euler space such that every interior point in it represents a physically distinct misorientation.

Asymmetric tilt grain boundary, ATGB. GB having the axis of misorientation parallel to the GB plane and the indices of the plane different in both grains.

Brandon criterion ν_m. A widely used relationship which gives the maximum angular deviation from a CSL.

Coincident axial direction/planar matching, CAD/PM. GB which is characterised by the near parallelism of the same planes from each neighbouring grain.

Coincidence site lattice, CSL. Specific misorientations which result in the coincidence of lattice points from each neighbouring grain.

Commensurate grain boundary. GB where the ratio of the planar GB unit cells is an integer.

Constrained coincidence, see **Near coincidence.**

Degrees of freedom, 'external' or 'extrinsic'. Relationship between the macroscopic degrees of freedom and some fixed external axes such as the axes which describe the specimen geometry.

Degrees of freedom, macroscopic. The geometry which relates the overall orientation change which occurs across a GB plane between two grains.

Degrees of freedom, microscopic. Small (< 1 atomic diameter) movements of two neighbouring grains parallel and perpendicular to the GB surface.

Disorientation θ_{min}/UVW. Angle/axis which contains, of the 24 symmetry-related solutions, the lowest angle.

Electron back-scatter diffraction, EBSD. A technique with a resolution of 200 nm for measurement-of individual orientations and thus GB geometry in a

scanning electron microscope.
Euler angles ϕ_1, Φ, ϕ_2. Description of a rotation (orientation or misorientation) by degeneration into three sequential rotations through angles (the Euler angles) about specific axes.
Euler space. A space in which misorientations (or orientations) are displayed when they expressed as Euler angles.
Favoured grain boundary. GB composed of only one type of structural unit.
General (random) grain boundary. Grain boundary which does not have a crystallographically ordered (periodic) structure.
'Geometrically special' grain boundary. GB which can be described by a periodic geometry.
Gnomonic projection. Projection of the reference sphere where the projection point is its centre and and the projection plane is tangential to its north pole.
Grain boundary, GB. The surface where two dissimilarly oriented crystals meet.
Grain boundary design (engineering). The concept and practice of deliberately increasing the proportion of special GBs in a material so that its overall properties will be improved.
Grain boundary geometry. A mathematical description of the macroscopic parameters which describe the crystallographic relationship between two neighbouring grains.
Grain boundary inclination. The angle between the GB plane and either the specimen surface normal, ϕ or the specimen surface, ϕ'.
Grain boundary plane. The surface between two grains which defines the GB, referred to by the crystallographic coordinates of its normal in both grains, N_1, N_2.
Grain boundary structure. Atomic level features of a GB, e.g. atomic positions, defect structure, chemistry.
Grain misorientation texture, see **Misorientation axis distribution.**
***I*-Line**. Triple grain junction where the dislocation content is balanced.
Intercrystalline structure distribution function, ISDF. A distribution which describes GB geometry in terms of five independent parameters relative to external axes.
Interface plane scheme. Scheme for describing GB geometry in terms of the indices of the GB in each grain plus a twist angle.
Low angle grain boundary (LA). GB which has a misorientation of $< 15°$.
'Mackenzie' triangle. Distribution of misorientation axes, displayed in a stereographic unit triangle, for the statistically random case.
Macrotexture. Averaged texture of a polycrystal measured by conventional means, e.g. x-rays.
Mesostructure. A well defined coherence structure (see **Orientation coherence**) between small clusters of grains.
Mesotexture. The texture between grains, i.e. the texture at grain boundaries.
Microtexture. A set of individual orientations from a polycrystal.

Misorientation. The re-expression of the orientation of two neighbouring grains in terms of the relative orientation between them.
Misorientation axis distribution. A representation of misorientation axes plotted relative to the crystal or specimen axes.
Misorientation distribution function, MODF. A probability based method for describing a distribution of misorientations. The **Physical MODF/Spatial orientation correlation/Measured MODF** takes into account the real spatial relationships between grains; the **Uncorrelated MODF/Textural or Statistical orientation correlation/Theoretical MODF** assumes that the input grains are randomly distributed in space.
Misorientation mapping/imaging. Combined representation of GB geometrical data and an image or diagram of the microstructure.
Misorientation matrix, M. 3×3 transformation matrix between the crystal axes of two neighbouring grains.
Multiple grain junction. A line where three or more GBs meet in a polycrystal.
Near coincidence site lattice. Extension of the range of the coincidence site lattice to some crystal systems other than cubic.
***O*-Lattice**. Array of all coinciding points, of which lattice points are a subset, between two misoriented lattices.
Orientation coherence (function), OCF. The preference for grains of a specified orientation to reside near other grains of another specified orientation.
Orientation correlation (function), $O(\delta g)$. The difference between a physical and a theoretical MODF.
Orientation matrix. 3×3 transformation matrix between the crystal axes of a grain and some fixed reference axes in the specimen.
Planar coincidence site density, PCSD. The density of coincidence sites in the GB plane.
Primary intrinsic grain boundary dislocations. Dislocations which comprise a low angle GB, i.e. which conserve the structure of the lattice.
Ranganathan's generating function. A mathematical function which gives all values of Σ and misorientation angle for a chosen misorientation axis.
Relative orientation, see **Misorientation.**
Rigid body translation. Atomic relaxations at the GB.
Rodrigues–Frank space, Fundamental zone of (RF space). The space in which all possible Rodrigues vectors reside for lowest angle solutions of the (mis)orientation.
Rodrigues vector, R-vector. A vector which defines a (mis)orientation. The direction and magnitude of the vector is a function of the (mis)orientation axis and angle respectively.
Rotation matrix, see **Orientation matrix.**
Rotation parameter space, see **Stereogram-based space.**
Rotation quaternion. Four integers which when squared then summed give Σ.
Secondary intrinsic grain boundary dislocations. Dislocation arrays which conserve the structure of a coincidence site lattice for small angular deviations

from perfect coincidence.

Σ-value. The ratio of the volume of the coincidence site lattice unit cell to the volume of the lattice cell.

'Special' grain boundary. Grain boundary which exhibits behaviour or has values for specific parameters which are very different (better in terms of material properties) than average.

Stereogram. Crystal plane normals or crystallographic directions depicted by stereographic projection.

Stereographic projection. Projection in which directions from an origin are projected from their intersection with a sphere centred on that origin to either the South or North pole. The point of intersection with the equatorial plane then depicts that direction.

Stereogram-based space. Space based on the stereographic projection, used in a two-dimensional form for displaying misorientation axes and GB plane normals, and in a three-dimensional form for displaying misorientations and MODFs.

Symmetrical tilt grain boundary, STGB. GB having the axis of misorientation parallel to the GB plane and the indices of the plane the same in both grains.

Symmetry-related misorientations. 24 equivalent solutions for a misorientation based on the symmetry of the cubic system.

Triple grain junction. Line where three GBs meet.

Twin plane. GB plane where every site in it is a coincidence site.

Twist grain boundary. GB having the axis of misorientation perpendicular to the GB plane and the indices of the plane the same in both grains.

***U*-Line**. Triple grain junction where the dislocation content is not balanced.

Unfavoured grain boundary. GB composed of more than one type of structural unit.

1

INTRODUCTION

1.1 SIGNIFICANCE OF GRAIN BOUNDARIES IN POLYCRYSTALS

The surface where two dissimilarly oriented crystals (grains) meet constitutes a grain boundary; thus at the most simple level a grain boundary (GB) is a crystallographic discontinuity. The GB width is on average less than two atomic diameters, which is sufficiently small for attractive forces to act across it, therefore the component grains in polycrystals remain united across GBs. However, GBs are far from being just inert surfaces which demarcate changes in orientation. The following important phenomena are influenced by GBs:

1. Phenomena directly resulting from movement (migration, sliding) of the GB itself (Aust and Rutter, 1959), e.g. recrystallisation (Berger *et al,* 1988), grain growth (Randle and Ralph, 1988a), creep (Ralph, 1980);
2. Transport phenomena e.g. solute segregation (Bouchet and Priester, 1987) or corrosion (Palumbo and Aust, 1990a);
2. Chemical reactions e.g. precipitation (Ainsley *et al,* 1979) and other phase transformations (Harase *et al,* 1990);
3. Mechanical properties e.g. strength and toughness (Wyrzykowski and Grabski, 1986; Lim and Watanbe, 1990);
4. Electrical properties (Nakamichi, 1990) e.g. with respect to dopants in semiconductors (Maurice *et al,* 1985);
5. Magnetic properties e.g. 'magnetic annealing' (Watanabe *et al,* 1990).

In other words GBs are *active structural elements* in crystalline materials, including some polymeric materials (Martin and Thomas, 1991). To illustrate this point further it is only necessary to compare the physical, chemical and mechanical properties of a polycrystalline material with those of the same material in single crystal form. The most well-known example is that of a single crystal turbine blade which performs far better under stress at high temperatures than its polycrystalline counterpart because grain boundary sliding (creep) is eliminated (Ralph, 1980, 1988).

It follows logically from the recognition of GBs as structural components that in general the behaviour of polycrystals cannot be considered to depend only upon the properties of an aggregate of individual grains; the macroscopic properties of polycrystals are governed by both grains *and* grain boundaries (Kurzydlowski, 1990). It is only recently that the relative importance, relationship and possible synergy between these two microstructural elements has been considered and has led to the proposal of a more physically relevant length scale in polycrystals which consists of grain clusters having a particular type of GB in common (Nichols *et al,* 1991a,b). There are methods available for the quantification of this clustering, also known as a 'mesostructure' (Adams *et al,* 1987), which measure the amount of 'orientation coherence' (Zhao, Adams and Morris, 1988).

The importance of GBs in the parent polycrystal can be further illustrated by the special cases of nanocrystalline materials and polycrystals which are essentially two-dimensional. Nanocrystalline materials typically have a grain size of 10nm or less (Gleiter, 1985). Hence they approach the upper bound of the relative proportions of grains and GBs in the material; for a grain size of 2nm the volume fraction occupied by grains and boundaries is equivalent, i.e. 50% each compared to a 0.3% volume fraction of GBs for a grain size of 1μm (Palumbo *et al,* 1990). The large grain boundary component modifies material properties, for example with respect to saturation magnetisation of α-iron. 'Two-dimensional' polycrystals could be either deposited on a substrate as a thin film (Grovenor *et al,* 1984) or rapidly solidified as a ribbon (Watanabe *et al,* 1989). Their properties are modified with respect to bulk polycrystals. One reason for this is the major role of surface effects when the GBs extend from the top to the bottom of the material.

1.2 GRAIN BOUNDARY STRUCTURE AND 'SPECIAL' BOUNDARIES

The principal physical feature which characterises the structure of a GB is its porosity or excess free volume (Qian *et al,* 1987) and associated short and long range elastic strain fields (Sutton and Vitek, 1983) compared to the lattice. Figure 1.1 is a simple schematic model for a GB which shows the 'width' of the GB region and some adsorbed species (segregants) in the GB. It is principally this excess free volume and the associated stress fields which confer upon GBs properties which are different from the lattice. For example, GBs have a greater propensity for segregation, diffusion, strain and defect accommodation, and various kinds of nucleation phenomena compared to the lattice. Values for specific parameters may vary by up to an order of magnitude or more. For example, figure 1.2 summarises comparisons between lattice and GB self diffusion coefficients (Gust *et al,* 1985).

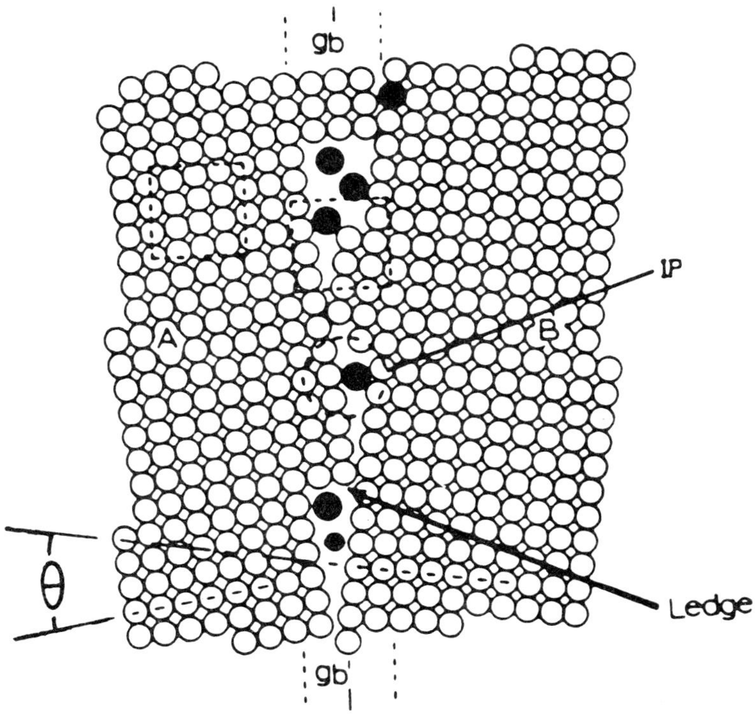

Figure 1.1 Schematic diagram of a GB between grains A and B where atoms are denoted by spheres. The dark spheres are segregant elements which can act as incipient precipitates (IP). Θ is the angle of misorientation between common directions in grains A and B (see chapter 2) (Murr *et al*, 1990).

There is a further division between classes of GBs themselves, rather than between GBs and the lattice, based on their properties. Frequently in the literature this division is denoted by describing boundaries as either general or special (Priester, 1989). General (also called random) boundaries are characterised by having average values for specific parameters such as diffusivity, mobility, energy, etc. By contrast special boundaries exhibit behaviour or have values for specific parameters which are very different to the average ones associated with random boundaries. These differences are directly related to the *structure* of the GB, and one important factor which confers special properties at certain GBs is particularly low excess free volume.

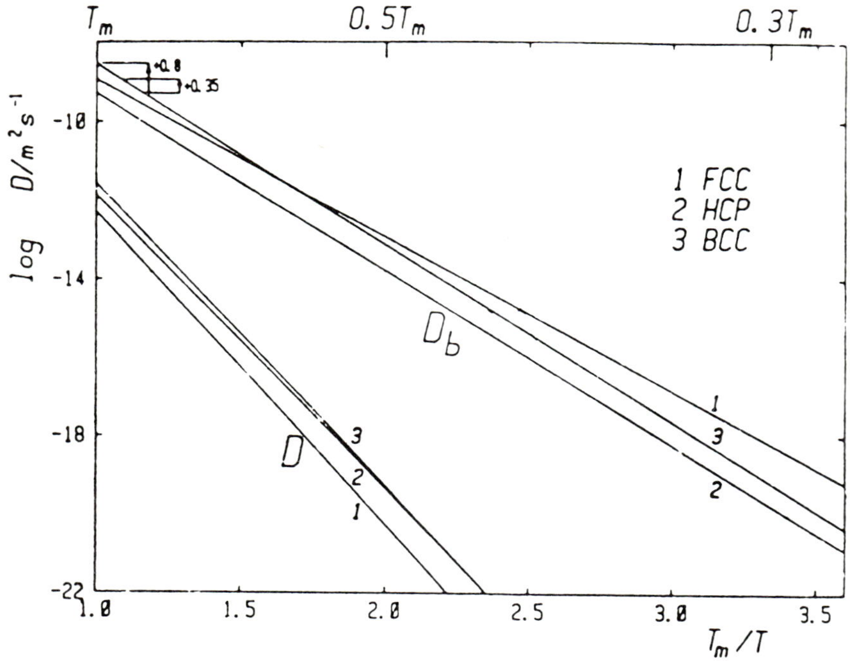

Figure 1.2 Graph showing a comparison of GB self-diffusion D_b and lattice self-diffusion D for a range of homologous temperatures T_m/T (Gust *et al,* 1985).

The significance of special GBs is that their properties are usually beneficial to the overall properties of the polycrystal, or could be exploited to be so. Appreciation of this fact has led to a great deal of research activity to elucidate precisely the structure of 'special' GBs and how the proportion of GBs with special properties could be maximised. The term 'grain boundary engineering (or design)' has been used in this context (Watanabe, 1984, 1988). For example, a fairly well studied area is how GB cavitation may be improved by manipulation of GB statistics (Lim and Raj, 1984a; Palumbo and Aust, 1990a; Field and Adams, 1992). However, the complete relationship between GB structure and properties is far from clear, and its clarification is one of the goals of research on GBs.

The term 'GB structure' implies a detailed knowledge of the atomic positions contributed by each grain at the GB (as in figure 1.1) in addition to the defect structure, chemistry, etc. Measurement of these parameters is only feasible on a small scale. Consequently studies of GBs in polycrystals is usually considered in terms of the overall GB geometry based on the crystallographic relationships between abutting grains rather than a detailed description of actual atomic positions at the interface and their interactions. This simplified geometrical approach is justifiable because the crystallographic relationship between abutting grains at a GB contributes the major part of the structure. Furthermore, and this is an important point, if statistically significant quantities of data are to be generated, crystallographic parameters are far more experimentally accessible and interpretable than those which involve invesigations on an atomic level. This is particularly true in the light of experimental techniques which have been developed in recent years for rapid crystallographic analysis such as electron back-scatter diffraction (EBSD) in a scanning electron microscope (SEM) (Dingley, 1981; Dingley and Randle, 1992; Randle 1992c). The purpose of this book is to describe in detail the whole approach to large-scale investigations of GB geometry, from theoretical aspects through data measurement, processing, representation, display, interpretation and application.

1.3 DEVELOPMENT OF TECHNIQUES FOR MEASUREMENT AND ANALYSIS OF GRAIN BOUNDARY GEOMETRY

The input data for the measurement of GB geometry is a combination of crystallographic and spatial information. In other words we need to know how the grains in a polycrystal fit together, i.e. which are contiguous, to be able to compute GB geometries. Furthermore the interpretation of these data is enhanced if we have other information about the local environment. Hence the essential experimental requirements are to be able to perform diffraction from selected regions which are less than the grain size and (usually) to be able to image these regions of the microstructure. To measure the orientation of the GB plane itself in addition to the misorientation requires additional spatial information. The most suitable technique for analysis of GB geometry is one which utilises electron microscopy so that diffraction and imaging can both be carried out *in situ*.

Until a few years ago, investigations which concerned the geometry of GBs were performed in the transmission electron microscope (TEM) but usually restricted to experiments on specially fabricated bicrystals (Goodhew *et al*, 1978; Forwood and Clarebrough, 1992). Few experiments were performed with the objective of collecting statistically significant quantities of data from polycrystals due to the labour intensity of the data acquisition. This was because the TEM route involved the somewhat tedious preparation of thin foil specimens, which (unless the grain size was very small) only allowed a small number of GBs to be analysed per foil. An alternative route, selected area channelling (SAC) in the scanning electron microscope (SEM), could be used on bulk specimens (Joy, 1975). However, specimen preparation was also difficult for this technique and the spatial resolution was about 10μm, which precluded application to very small grains or subgrains.

In the last few years there have been great improvements both in the experimental techniques available and the software for on-line interrogation of diffraction patterns. Currently, the raw data for GB geometrical analyses can be measured and processed on-line semi-automatically both in the TEM and in the SEM (Schwarzer, 1990). The principal technique for crystallographic analysis in the SEM, which has been implemented during the last decade and is now superceding SAC, is electron back-scatter diffraction (EBSD). This technique has a spatial resolution of 200nm, is performed on specimens prepared by simple metallographic techniques and allows particularly rapid and convenient collection and analysis of data. It is largely because of developments in experimental techniques that the study of GB geometries in polycrystals, and their subsequent relationship with the properties of both the constituent grains and the GBs themselves, has been able to broaden into an area of major research activity.

In addition to the instrumental and software developments to measure GB geometry which were mentioned in section 1.2, the methods by which GB geometrical parameters can be manipulated and represented are becoming increasingly sophisticated (Field *et al,* 1991). This is in response to the need to analyse and interpret the large quantities of data which can now be generated, which is a result of increased research activity in this area. The task of presentation of GB geometrical data is complex because frequently several statistical parameters as well as spatial information need to be incorporated into the data output. In other words both the local environment and crystallography of individual GBs have been probed concurrently. Furthermore, for a more complete description of the microstructure, these data can be united with parallel information - crystallographic and spatial - about the constituent grains themselves in addition to the GBs.

1.4 SCOPE OF THIS BOOK

The aim of this book is to provide a comprehensive guide to all the stages of data collection and analysis connected with GB geometry in cubic polycrystals. It will be of particular value to the experimentalist or anyone who requires a state-of-the-art overview of the field of GB geometry research. The title of the book refers specifically to GB *geometry*: there are other aspects of GB structure and structure/property relationships which are not covered in detail in this book but nonetheless contribute information about GB structure. These topics include high resolution electron microscopy (HREM) (Merkle and Smith, 1987), studies of dislocation interactions with GBs including dislocation spreading/dissociation (Swiatnicki *et al,* 1986), modelling or simulation of GB structure (Pond and Vitek, 1977; Balluffi *et al,* 1987), the measurement of any

GB properties such as diffusion, energy, etc (McLean, 1973; Balluffi, 1982) and GB geometry in non-cubic materials (Bleris *et al,* 1982; Grimmer, 1989).

The central theme of this book is the collection and analysis of orientation data relating both to adjoining grains and to the plane of the GB itself. Measurement of both of these parameters is accessible in the SEM. However, where a finer probe size is required than can be generated in the SEM, e.g. for subgrains or cold-worked structures, the same parameters can be obtained in the TEM. The important point to be made here is that measurement of the GB geometrical parameters is rapid (especially for the misorientation), accurate and does not require extremely specialised equipment or skills, yet can provide valuable information about GB structure. This information can, if required, be combined with other microstructural data, e.g. that concerning properties. More detailed studies of GBs would include mapping of the atomic positions at the boundary, defect structure analysis, chemical species profiles and computer simulations; however, resources and labour intensity render such studies unfeasible except on a small scale. By contrast, stastistically significant quantities of GB geometrical data can be generated readily using modern experimental techniques and methodologies. Hence the geometrical route offers the optimum compromise between advances in understanding and practicability for the study of GBs and their role in material properties control.

The rest of this book is arranged as follows. In chapter 2 the theoretical aspects of GB geometry are introduced and explained. This includes the terminology and mathematical formulations associated with a GB for the general case. This theoretical approach is extended in chapter 3 to GBs which have non-random geometries. In chapter 4 the emphasis is switched from the theoretical framework to the actual collection of data, and the principles of this are described in detail, with particular emphasis on EBSD because this is a relatively new technique. Chapter 5 then moves on to how the raw data, once it has been acquired, is processed to produce a measure of the GB geometry. The algorithms which can be used are described fully with examples. The next step in the sequence of GB geometry analysis is to have a means of representing and displaying the data in such a way that it can be interpreted. There are several routes for data representation and these are explored in chapter 6. Finally in chapter 7 all the topics discussed in the previous chapters are brought together by means of a résumé of actual investigations of GB geometry taken from the literature.

2

THEORETICAL ASPECTS OF GRAIN BOUNDARY GEOMETRY I: GENERAL BOUNDARIES

2.1 INTRODUCTION

Very early models of GBs proposed that their structure was equivalent to an amorphous intergranular glue (Rosenhain and Humphrey, 1913). At that time no recognition was given to interfacial atomic structure. The later 'transitional lattice' model (Hargreaves and Hill, 1929) and a refinement of this, the 'island' model (Mott, 1950), envisaged that atoms in a GB were transitional between the abutting crystal lattice sites. Two major advances towards 'modern' views of GB structure occurred when dislocation theories were incorporated into models (Read and Shockley, 1950), and the crystallographic (that is, the orientational) parameters of the interfacing grains were considered (Kronberg and Wilson, 1949). The latter led to the concept that certain GBs were composed of periodic arrays of atoms from each grain (Brandon *et al,* 1964; Brandon, 1966) which provided a physical reason for observations which had been made that certain boundaries had properties which were different from average. These two models - dislocation structure of GBs and periodicity, i.e. coincidence at boundaries - became fused in the most sophisticated and elegant approach to GB geometry, the O-lattice model (Bollmann, 1970). This very short summary of the historical background to present-day understanding of GB geometry is amplified in reviews elsewhere (McLean, 1957; Chadwick and Smith, 1976; Gleiter 1982).

The purpose of this chapter is to introduce and define the theoretical bases and terms which are fundamental to descriptions of GB geometry. Note again that the term 'GB geometry' is used to denote *crystallographic* parameters, that is, how the orientation of grains residing in the same polycrystal are related. Additionally the relationship between these parameters and some external reference axes can be included in the data. It also needs to be re-emphasised that GB geometrical data are experimentally accessible through standard procedures, and it is the main function of this book to describe these procedures. The term 'GB structure' on the other hand is concerned with atomic positions in the GB region and requires more sophisticated methods for its investigation. In this book the coincidence approach to GB geometry, known fully as the coincident site lattice model, CSL, (see section 3.2) is described thoroughly because, as shown in chapter 7, it is the main tool in current use to classify GB geometry in polycrystals. Alternative descriptions of grain boundary structure, such as the O-lattice model and the 'near CSL' model which extends the CSL to non-cubic systems, are described briefly (see section 3.3).

2.2 GRAIN BOUNDARY DEGREES OF FREEDOM

The starting point for a geometrical description of a GB is a definition of its degrees of freedom. A GB in a bicrystal has eight degrees of freedom in total; five of these are known as macroscopic, and the other three microscopic (Goux, 1974; Wolf, 1990a). Of the five macroscopic degrees of freedom, four define two directions (two each) and one defines an angle; (a direction or a plane normal is specified by *three* direction cosines but only *two* are independent variables - see equation 2.5). The identity of the two directions and angle depends upon whether the interface-plane scheme or the misorientation scheme is used to describe the GB geometry. These two schemes are described in the next two subsections. The three microscopic degrees of freedom refer to translations which are mutual movements of the bicrystal halves parallel and perpendicular to the GB surface. These translations occur on the atomic level and act so as to minimise the GB free energy. They are difficult either to measure or manipulate; thus only the macroscopic degrees of freedom fall within the scope of this book.

The macroscopic degrees of freedom of a GB characterise the geometry which relates the overall orientation change occurring across the grain boundary plane in a bicrystal. This definition emphasises the crystallographic nature of GB geometry; essentially it refers to the orientation relationship, which is usually called the misorientation, across the surface of the GB. The surface of the GB is referred to as the GB *plane* even though in reality it may not always be planar (Lange, 1967). The orientation of each grain is defined by the orientation of its crystal coordinate system 100, 010, 001 with respect to a fixed reference system (see section 2.2.3).

2.2.1 The interface-plane scheme

The crystallographic relationship across a GB plane is illustrated in figure 2.1. This figure shows schematically the sets of lattice planes from each grain which are parallel to the GB. The orientations of the interfacing grains are denoted in figure 2.1 with respect to their principal coordinate axes $x_1y_1z_1$ and $x_2y_2z_2$ which are the crystal axes 100, 010, 001 for each grain. The lattice plane stacks which abut the GB are labelled by the normals to these planes, N_1 and N_2, which are indexed in the coordinate systems $x_1y_1z_1$ and $x_2y_2z_2$ respectively. In this book we will adopt the convention of referring to the crystallographic orientation of the GB plane by the direction of the normal to it. This avoids possible confusion by restricting the term 'orientation' for application to the three-dimensional case, i.e. for grains.

In figure 2.1a it is clear that N_1 and N_2, the normals to the lower (1) and upper (2) grains respectively, are parallel. In a completely general bicrystal grain 2 is twisted about N_2 (= N_1) with respect to grain 1 as shown in figure 2.1b. This twist angle is denoted φ. For the case illustrated schematically in figure 2.1a, $\varphi = 0$.

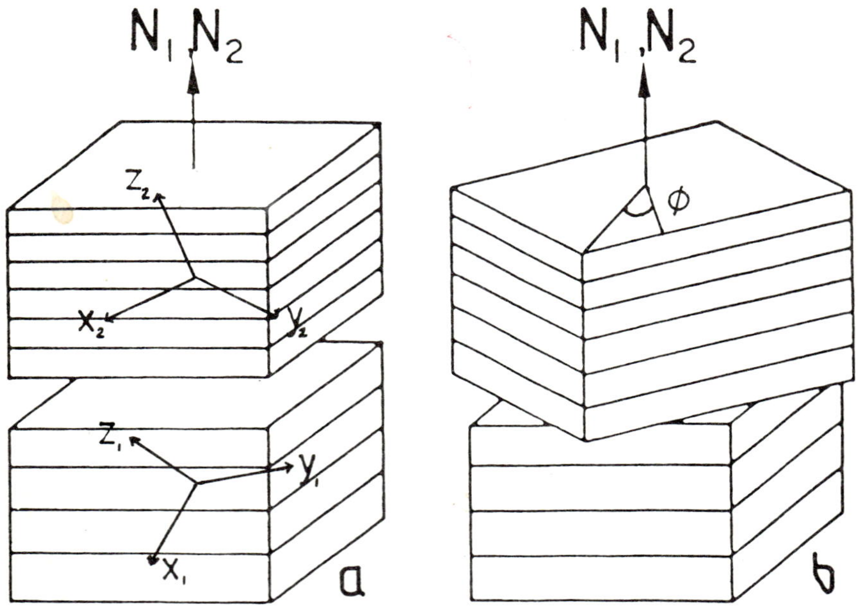

Figure 2.1 Schematic illustration of the lattice planes, with normals N_1 and N_2, either side of a GB between grains 1 and 2. In (b) the planes are related by a twist angle φ, and in (a) $\varphi = 0$. (Wolf and Lutsko, 1989).

In the geometrical description of a general GB described above and shown in figure 2.1b the five degrees of freedom are two from the GB plane normal in each of the two grains and one from the twist angle. This description of GB geometry is known as the interface plane scheme (Wolf and Lutsko, 1989). It is pertinent because it focusses on the concept of a GB as a joining together of two surfaces, followed in general by a twist rotation, to form a bicrystal. The interface plane scheme description of GB geometry is visually direct and, as will be discussed in section 2.4 and 3.2.5, emphasises the symmetry properties of GBs.

2.2.2 The misorientation scheme

An alternative interpretation of the GB degrees of freedom does not start from the concept of the joining of two GB planes as in the interface-plane scheme, but from the standpoint of the relative rotation between the orientations of the two neighbouring lattices. This leads to the angle/axis notation to describe the relative rotation (misorientation) between two grains. The axis/angle notation is used more commonly than the interface-plane notation.

In order to understand the fundamental idea of a misorientation it is useful to start by imagining that the two lattices can interpenetate, which is shown schematically in figure 2.2a. Although the position of the surface which constitutes the actual GB itself is shown on figure 2.2a, for the moment we are focussing solely on how the two lattices are related. As we will see in chapter 3, the concept of lattice misorientation is central to the development of the CSL model.

The rotation of any rigid body in space relative to some fixed system is accomplished by identifying a direction (axis) such that a rotation about this axis through a specific angle results in the alignment of the body with the fixed reference system. For the GB case the reference system is the crystal axes of the first grain in a pair of neighbouring grains, where the designation of 'first' and 'second' grain is arbitrary. The crystal axes of the second grain are aligned with those of the first by a rotation through the angle of misorientation, Θ, about the axis of misorientation, *UVW*, which in some texts is denoted by the symbol ℓ. Θ and *UVW* are marked on figure 2.2a. Note that Θ is not equivalent to φ in the interface-plane scheme (figure 2.1).

Figure 2.2a is in fact a special case of GB geometry called an asymmetrical tilt GB because *UVW* lies in the GB plane (see section 3.2.5). The general case, where *UVW* is related in an arbitrary way to the GB plane, is shown in figure 2.2b. However, it is easier to view and understand the concept of a relative rotation between interpenetrating lattices by reference to figure 2.2a rather than the general case in figure 2.2b.

A particularly graphic way to illustrate Θ/UVW is by stereographic projection (Randle and Ralph, 1986). Figure 2.3 shows that the crystal axes of grain 2 are superimposed onto those of grain 1 by a rotation through Θ, which is 60° in this case, about *UVW*, which is 111. (60°/111 is a CSL with $\Sigma = 3$: this notation is explained in section 3.2). From figure 2.3a it can be appreciated that a defining condition for *UVW* is that it has the *same Miller indices* in the coordinate systems of both grains because it is equidistant from the pole pairs (100_1 and 100_2), (010_1 and 010_2), (001_1 and 001_2). The axes of

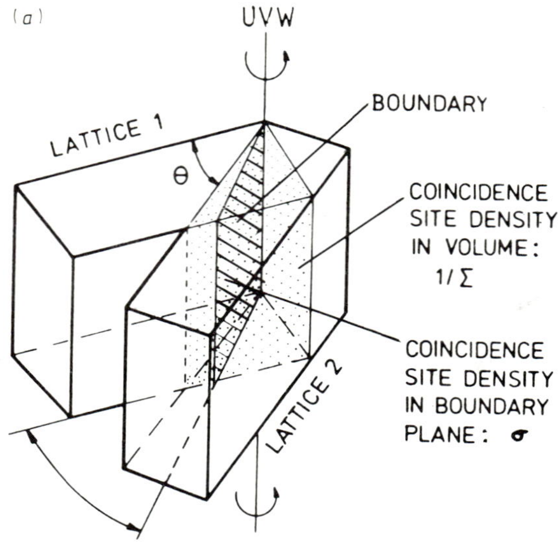

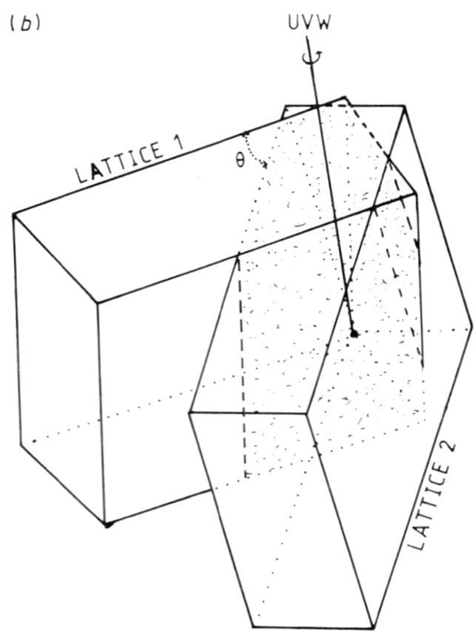

Figure 2.2 Illustration of the interpenetration of two lattices. The relationship between the lattices is described by a rotation through the angle of misorientation, Θ about the axis of misorientation, *UVW*. (a) The position of the GB itself within the interpenetrating lattices is marked. For this non-general case *UVW* is parallel to the GB, which is thus an asymmetrical tilt GB. (Fischmeister, 1985). (b) Illustration of interpenetrating lattices for the general case, i.e. a non-special GB.

the stereogram are the crystal axes of grain 1, the reference grain. Figure 2.3b shows the crystal axes of grain 2 re-indexed with reference to grain 1, so that now both grains share a common reference system. This figure is referred to again in section 2.3.1. and 3.2.

The axis of misorientation represents two degrees of freedom and the misorientation angle one. The angle/axis description alone does not specify fully the geometry of a GB because not all five degrees of freedom are used. The angle/axis defines only the relationship between the two neighbouring lattices and not the position of the GB itself. A simple schematic illustration of how the position of the GB plane is related to the orientation of each lattice is shown as a two-dimensional projection of two misoriented lattices in figure 2.4. First, two lattices are 'notionally interpenetrated' (figure 2.4a). The position of the GB is marked in figure 2.4a, and for convenience it is perpendicular to the plane of the page. The next step in the construction of a GB within the two interpenetrating lattices is shown in figure 2.4b. That part of the grain 1 lattice which is on the right-hand side of the boundary and that part of the grain 2 lattice which is on the left-hand side of the boundary are discarded, because in real, physical space an interpenetrating lattice cannot exist: it is simply a device to construct the GB geometry. Finally, in figure 2.4c the two lattices are shown having undergone a mutual 'rigid body translation' (described by the microscopic degrees of freedom) such that the best 'fit' is obtained at the GB.

Figure 2.4 thus shows how the actual position of the GB is incorporated with the misorientation geometry, which essentially describes only the relationship between two lattices. Similarly to figures 2.2a and b, it is only for convenience of representation on the printed page that the GB is chosen to be perpendicular to the page in figure 2.4; in reality the GB could be inclined at any angle within the interpenetrating lattice. Here Θ/UVW is 38.9°/<011>, N_1 is 111 and N_2 is 511. (This is a $\Sigma = 9$ CSL - see section 3.2).

The position of the GB surface, as defined by the coordinates of its normal in *one* of the grains, N_1 or N_2, takes up the last two degrees of freedom in the angle/axis (i.e. misorientation) description of GB geometry. Clearly for a specific Θ/UVW N_1 can take any value because the GB can in principle reside anywhere within the interpenetrating lattices. To supply the indices of N_2 in addition to those of N_1 overdefines the boundary geometry and in section 2.4 we see how N_2 can be obtained from Θ/UVW and N_1. Implicit in the misorientation scheme description of the five degrees of freedom is that a rotation of Θ about UVW superimposes N_2 upon N_1. This rotation is shown on a stereogram in figure 2.5 for an example where Θ/UVW is 70.5°/110, N_1 is $\bar{1}15$ and N_2 is $1\bar{1}1$.

The angle/axis is the most commonly used descriptor of grain boundary geometry principally because it is both easy to measure and it relates directly to data analysis methods such as CSL classification and representation in Rodrigues-Frank space (see section 6.3). Furthermore, the angle and axis of

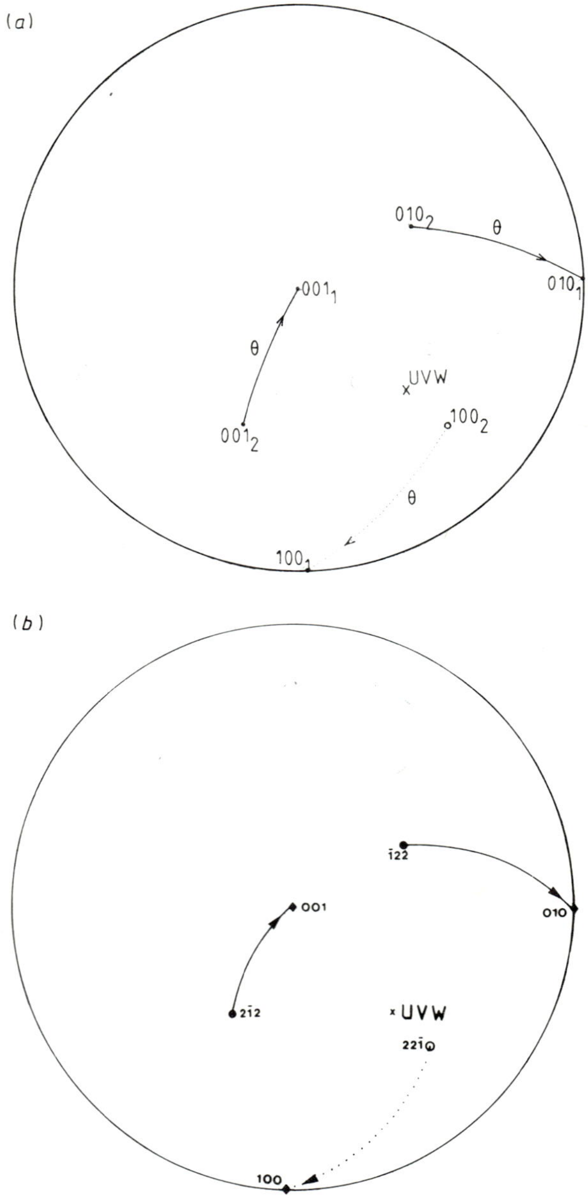

Figure 2.3 (a) Stereogram illustrating the angle/axis of misorientation concept. The axes of the stereogram are the crystal axes of grain 1. The crystal axes of grain 2 are superimposed on those of grain 1 by a rotation Θ through *UVW*. In (b) the misorientation is the same, but the crystal axes have been rotated such that both grains now share a common reference system *XYZ*.

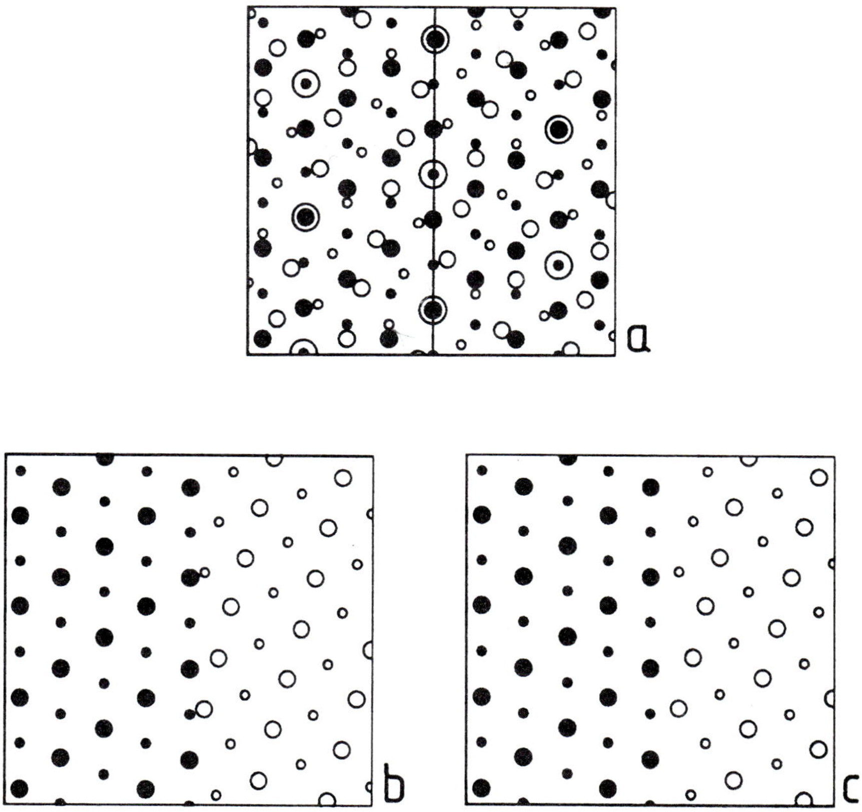

Figure 2.4 (a) Two interpenetrating lattices viewed along the axis of misorientation, $0\bar{1}1$. Atoms from the left- and right-hand grains are denoted by black and white symbols respectively, and two atomic layers are distinguished by different sized symbols. (b) black and white atoms have been removed on opposite sides of the GB. (c) rigid body translation at the GB. Courtesy of K. L. Merkle (Merkle, 1989).

misorientation may be relevant as individual parameters, e.g. in the case of low angle GBs. Frequently, reported measurements of 'GB geometry' do not include the GB plane normal. Hence only three of the five degrees of freedom for the GB are determined.

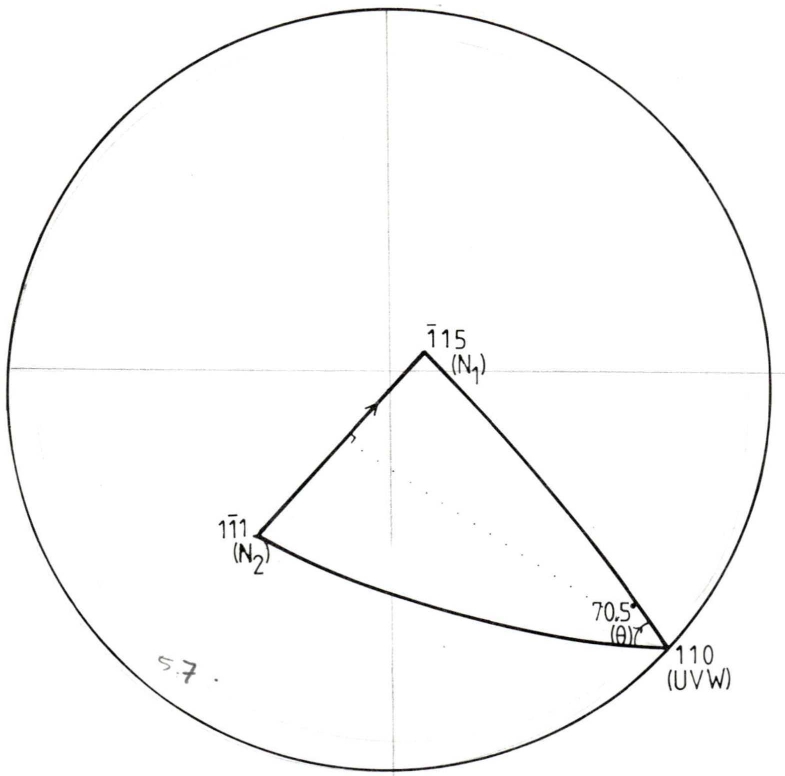

Figure 2.5 Stereogram illustrating the relationship between Θ, *UVW*, N_1 and N_2, indexed in the coordinate system of the reference grain, grain 1. For this example a rotation of 70.5° about 110 aligns N_1 and N_2, which are $1\bar{1}1$ and $\bar{1}15$ respectively.

To summarise the foregoing two subsections, the geometry of a GB (i.e. the five degrees of freedom) can be expressed in two ways:

$$\text{Interface-plane scheme} = N_1, N_2, \varphi \tag{2.1}$$

$$\text{Misorientation (+plane) scheme} = \Theta, UVW, N_1 \tag{2.2}$$

These two expressions are completely equivalent, and their mathematical relationship can be illustrated by decomposing the overall GB geometry into its tilt and twist components which is described in section 2.4.

2.2.3 'External' degrees of freedom

So far the degrees of freedom of a GB have been discussed in terms of the parameters required to define the orientation of one grain relative to its neighbour, and the reference axes are therefore the crystal axes of one of the grains. We may also wish to know how the misorientation and the GB plane normal are related to some fixed external axes, such as the axes which define the overall specimen geometry. These additional parameters can be considered to be 'external' or 'extrinsic' degrees of freedom.

The thermomechanical processes which are used to form and shape a material play a dominant role in selecting the orientations of grains in the polycrystal. This leads to a preponderance of certain orientations which is called preferred orientation or, more commonly, texture (Bunge, 1987). A recent extension to this long-established field of research is microtexture, which means individual, spatially specific orientation measurements (Randle, 1992c). Clearly the measurement of microtexture and GB geometry are interdependent because the orientations of neighbouring grains have to be measured in order to obtain the misorientation.

In (micro)texture determination the orientation of a grain is expressed relative to reference axes in the specimen. It follows that the misorientation may also be related to these specimen axes. A common example of specimen reference axes are the rolling direction (RD), normal to the rolling direction (ND) and the transverse direction (TD) in a rolled sheet product. These axes are orthogonal. Furthermore, the specification of the boundary plane may also be made with respect to the specimen axes in addition to the orientation of each grain. If these external degrees of freedom are included in the analysis of GB geometry, the total number of degrees of freedom is increased from five to eight (Field and Adams, 1992). The consequences of this are discussed further in section 6.5.

2.3 MATHEMATICAL CHARACTERISATION OF A MISORIENTATION

The angle/axis pair is one way in which the misorientation geometry of a GB may be described mathematically; there are other equivalent descriptions which will be introduced in this section. First, we consider some general comments about the form of *UVW*.

Usually, *UVW* is expressed in Miller indices or direction cosines, which are both Cartesian coordinates. An alternative form for *UVW* is two polar (spherical) coordinates as shown in figure 2.6. These are often denoted by (Θ,ψ) with the misorientation angle ω (Pospiech *et al*, 1986). Since there could be confusion with the more common usage of Θ as misorientation angle, here we will use (p_1,p_2) for *UVW* expressed as polar coordinates (figure 2.6) and p_3

for the misorientation angle. The connection between *UVW* expressed in both direction cosines $U_d V_d W_d$ and spherical coordinates (p_1, p_2) is

$$
\begin{aligned}
U_d &= \sin p_2 \cos p_1 \\
V_d &= \sin p_2 \sin p_1 \\
W_d &= \cos p_2
\end{aligned}
\tag{2.3}
$$

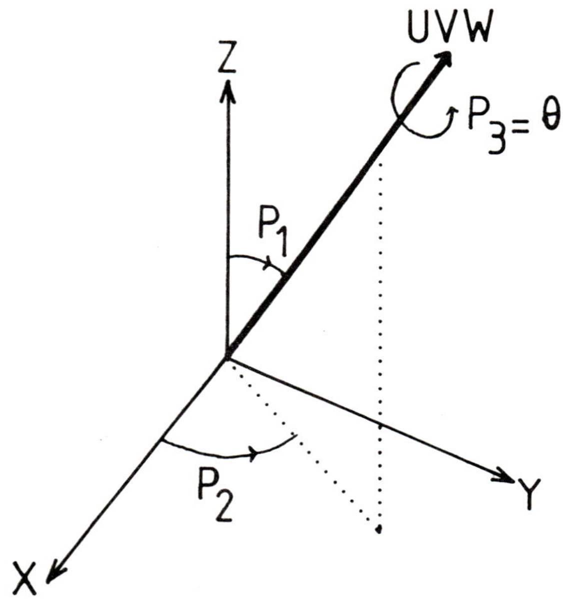

Figure 2.6 Diagram showing how the axis of misorientation can be expressed as polar coordinates, p_1, p_2 relative to orthogonal axes *XYZ*. p_3 is the angle of misorientation.

Also the connection between *UVW* in direction cosines and Miller indices $U_m V_m W_m$ is

$$
\begin{bmatrix} U_d \\ V_d \\ W_d \end{bmatrix} = C \begin{bmatrix} U_m \\ V_m \\ W_m \end{bmatrix} \tag{2.4}
$$

where $C = (U_m{}^2+V_m{}^2+W_m{}^2)^{1/2}$. Throughout this book UVW will be used to denote the components of the misorientation axis and the suffixes *d* or *m* will be omitted unless it is necessary to make a clear distinction between them.

Finally, $U_dV_dW_d$ is a unit vector and so the following condition applies:

$$U_d^2+V_d^2+W_d^2 = 1 \tag{2.5}$$

For example if *UVW* is 111 in Miller indices, its alternative notations are 0.577, 0.577, 0.577 as direction cosines and 54.7°, 45.0° as polar coordinates.

2.3.1 The misorientation (rotation) matrix

The rotation necessary to transform the crystal axes of grain 2 onto grain 1 can be expressed in the form of a 3x3 matrix, **M**:

$$\mathbf{M} = \begin{bmatrix} a_{11} & a_{12} & a_{13} \\ a_{21} & a_{22} & a_{23} \\ a_{31} & a_{32} & a_{33} \end{bmatrix} \tag{2.6}$$

The matrix columns are the direction cosines of the crystal axes of grain 2 referred to the coordinate system of grain 1, which is chosen here to be the reference grain. For the case illustrated in figure 2.3, and transforming from Miller indices to direction cosines, the misorientation matrix is

$$\mathbf{M} = \begin{bmatrix} .6667 & -.3333 & .6667 \\ .6667 & .6667 & -.3333 \\ -.3333 & .6667 & .6667 \end{bmatrix} \tag{2.7}$$

Thus the first column of the matrix gives the coordinates of the 100 direction in grain 2 with reference to grain 1, and similarly for the other two columns and the 010 and 001 directions (see figure 2.3b). A property of the misorientation matrix is that it is orthogonal: the sums of the squares of each row and each column are unity (see equation 2.5), and the dot product between each column and row vector is zero, i.e. the angle between them is 90°. Throughout this book the matrix **M** will be referred to as the misorientation matrix rather than the rotation matrix to emphasise the connection with the misorientation of two lattices which forms a GB. The term 'rotation matrix' will be reserved for reference to either the orientation of grains, or rotations in general.

Because of the relationships between columns and rows of **M**, it contains only three independent variables (degrees of freedom) although it is made up of nine numbers. The three independent variables can be expressed as the angle/axis pair as follows:

$$\cos \Theta = (a_{11}+a_{22}+a_{33}-1)/2 \tag{2.8}$$

$$U{:}V{:}W = a_{32}-a_{23}{:}a_{13}-a_{31}{:}a_{21}-a_{12} \tag{2.9}$$

If $\Theta = 180°$ then UVW is given by:

$$U{:}V{:}W = (a_{11}+1)^{\frac{1}{2}}{:}(a_{22}+1)^{\frac{1}{2}}{:}(a_{33}+1)^{\frac{1}{2}} \tag{2.10}$$

Conversely the elements of **M** in terms of Θ/UVW are given by:

$$\begin{aligned}
a_{11} &= U^2(1-\cos\Theta) + \cos\Theta \\
a_{12} &= UV(1-\cos\Theta) - W\sin\Theta \\
a_{13} &= UW(1-\cos\Theta) + V\sin\Theta \\
a_{21} &= VU(1-\cos\Theta) + W\sin\Theta \\
a_{22} &= V^2(1-\cos\Theta) + \cos\Theta \\
a_{23} &= VW(1-\cos\Theta) - U\sin\Theta \\
a_{31} &= WU(1-\cos\Theta) - V\sin\Theta \\
a_{32} &= WV(1-\cos\Theta) + U\sin\Theta \\
a_{33} &= W^2(1-\cos\Theta) + \cos\Theta
\end{aligned} \tag{2.11}$$

The advantage of the matrix description of misorientation is its mathematical manipulability, allowing rotations to be combined (see sections 2.3.2 and 3.4) and misorientation differences to be measured, e.g. for comparison with a CSL misorientation (see sections 3.2.2 and 3.2.3). Thus if

Θ, *UVW* and N_1 are known we can obtain N_2 from premultiplying it by the misorientation matrix formulated from Θ and *UVW* using equation 2.12:

$$N_2 = \mathbf{M}\, N_1 \tag{2.12}$$

Taking a general example, if Θ, *UVW* and N_1 are 34.7°, 0.767 0.628 0.131 and 0.924 0.267 0.259 respectively, N_2 is given by:

$$\begin{bmatrix} .957 \\ .277 \\ .020 \end{bmatrix} \begin{bmatrix} .927 & .011 & .375 \\ .160 & .892 & -.422 \\ -.340 & .451 & .825 \end{bmatrix} = \begin{bmatrix} .924 \\ .267 \\ .259 \end{bmatrix} \tag{2.13}$$

2.3.2 Euler angles

Any rotation may be degenerated into three sequential rotations through certain angles which are known as the Euler angles (Bunge, 1985). Using the nomenclature for a misorientation, the crystal axes for neighbouring grains 1 and 2 are specified as $x_1y_1z_1$ and $x_2y_2z_2$ with grain 1 in the reference position. To align the axes of grain 2 with those of grain 1, the following three rotations are performed in the order shown:

$$\text{rotation 1} = \mathbf{m_1} = \varphi_1 \text{ about } z_2; \tag{2.14a}$$

$$\text{rotation 2} = \mathbf{m_2} = \Phi \text{ about } x'_2; \tag{2.14b}$$

$$\text{rotation 3} = \mathbf{m_3} = \varphi_2 \text{ about } z_1 \tag{2.14c}$$

These rotations are illustrated on figure 2.7. The significance of the 'prime' symbol in equation 2.14b is that after the first rotation (in equation 2.14a) the original direction of x_2 has moved to x'_2. After the second rotation z_1 and z_2 are aligned, and so the third rotation is performed about the direction common to both of them. Some descriptions of the Euler angle representation of a misorientation use different symbols from those given above; for instance α about z_2, ß about y'_2, γ about z_1. α,ß,γ may also be referred to as ψ_c, Θ_c, Φ (Roe, 1965). Note also that for the Roe notation the second rotation is about y'_2, not x'_2. The total rotation is equivalent.

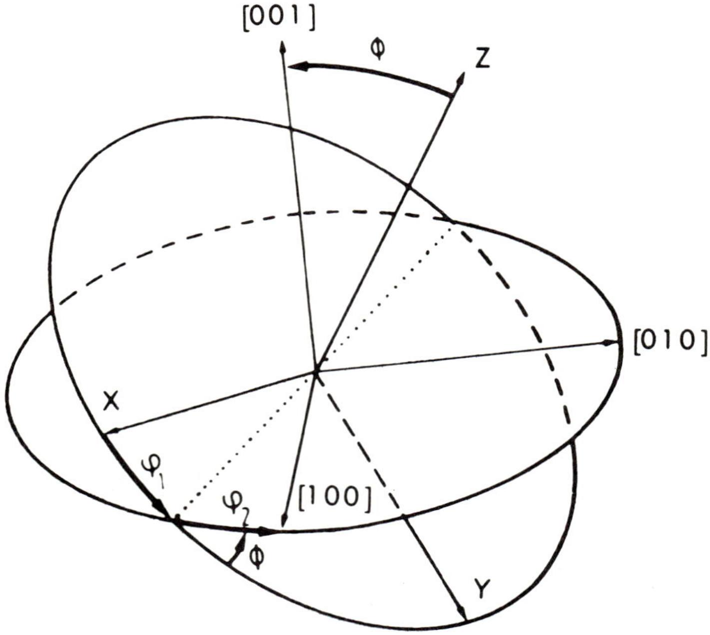

Figure 2.7 Definition of the Euler angles, φ_1,Φ,φ_2, which describe the rotation between two sets of axes, here denoted as *XYZ* and 001, 010, 100. (Adams *et al,* 1987)

Euler angles are well known as a means of displaying grain orientations, particularly continuous distributions, which are known as orientation distribution functions (ODFs) and displayed in Euler space (Bunge, 1985, 1987). In the same manner misorientation distributions can be displayed in Euler space. Such a distribution is known as the misorientation distribution function (MODF). These are discussed in detail in section 6.4.

Mathematically, the three rotations in equation 2.14 can be expressed in matrix form:

$$\mathbf{m}_1 = \begin{bmatrix} \cos\varphi_1 & \sin\varphi_1 & 0 \\ -\sin\varphi_1 & \cos\varphi_1 & 0 \\ 0 & 0 & 1 \end{bmatrix} \qquad (2.15a)$$

$$\mathbf{m}_2 = \begin{bmatrix} 1 & 0 & 0 \\ 0 & \cos\Phi & \sin\Phi \\ 0 & -\sin\Phi & \cos\Phi \end{bmatrix} \quad (2.15b)$$

$$\mathbf{m}_3 = \begin{bmatrix} \cos\varphi_2 & \sin\varphi_2 & 0 \\ -\sin\varphi_2 & \cos\varphi_2 & 0 \\ 0 & 0 & 1 \end{bmatrix} \quad (2.15c)$$

By matrix multiplication the rotation matrix, which for the GB case is the misorientation matrix, is obtained:

$$\mathbf{M}=\mathbf{m}_3\mathbf{m}_2\mathbf{m}_1 \quad (2.16)$$

From equation 2.16, the elements of **M** in terms of the Euler angles φ_1,Φ,φ_2 are:

$$\begin{aligned}
a_{11} &= \cos\varphi_1 \cos\varphi_2 - \sin\varphi_1 \sin\varphi_2 \cos\Phi \\
a_{12} &= \sin\varphi_1 \cos\varphi_2 + \cos\varphi_1 \sin\varphi_2 \cos\Phi \\
a_{13} &= \sin\varphi_2 \sin\Phi \\
a_{21} &= -\cos\varphi_1 \sin\varphi_2 - \sin\varphi_1 \cos\varphi_2 \cos\Phi \\
a_{22} &= -\sin\varphi_1 \sin\varphi_2 + \cos\varphi_1 \cos\varphi_2 \cos\Phi \\
a_{23} &= \cos\varphi_2 \sin\Phi \\
a_{31} &= \sin\varphi_1 \sin\Phi \\
a_{32} &= -\cos\varphi_1 \sin\Phi \\
a_{33} &= \cos\Phi
\end{aligned} \quad (2.17)$$

Taking the misorientation matrix in equation 2.7 as an example, the Euler angles obtained using equation 2.17 are:

$$\varphi_1 = 45.00°;\ \Phi = 70.53°;\ \varphi_2 = 45.00° \quad (2.18)$$

Equation 2.16 demonstrates the advantage of the matrix representation of a misorientation for combining several rotations by matrix multiplication.

2.4 SYMMETRY-RELATED MISORIENTATIONS

When a misorientation is generated the axes of the first grain are fixed with respect to the axes of the second grain. However, the symmetry of the cubic crystal system dictates that the axes of the second grain may be chosen in more than one way. For any crystal system the number of these equivalent axes are (Pumphrey and Bowkett, 1971):

$$n = 1 + n_2 + 2n_3 + 3n_4 + 5n_6 \qquad (2.19)$$

where n_x is the number of axes of order x. For the cubic system, this gives the identity, six two-fold axes, four three-fold axes, and three four-fold axes, giving a multiplicity of 24. This means that a misorientation in the cubic system can be described in 24 different - but equivalent - ways, which in turn gives 24 solutions of the angle/axis pair, the interface-plane or the Euler angle notation.

The symmetry-related solutions for a misorientation are generated by representing each symmetry operation as a matrix **T** and premultiplying the original misorientation matrix by each one in turn:

$$\mathbf{M}' = \mathbf{T}_i\mathbf{M} \qquad 2.20$$

where $i = 1,2,\ldots 24$. The 24 **T**-matrices are listed in Table 2.1, starting with the identity matrix, **I**:

$$\mathbf{I} = \begin{bmatrix} 1 & 0 & 0 \\ 0 & 1 & 0 \\ 0 & 0 & 1 \end{bmatrix} \qquad (2.21)$$

By convention, right-handed coordinate systems are used and a right-handed screw direction outward along the rotation axis is the positive rotation direction (Mykura, 1980).

TABLE 2.1

MATRICES REPRESENTING THE 24 SYMMETRY OPERATIONS FOR THE CUBIC SYSTEM

1 0 0	$\bar{1}$ 0 0	0 0 $\bar{1}$	$\bar{1}$ 0 0	0 0 1	1 0 0
0 1 0	0 $\bar{1}$ 0	0 1 0	0 1 0	0 1 0	0 0 $\bar{1}$
0 0 1	0 0 $\bar{1}$	1 0 0	0 0 $\bar{1}$	$\bar{1}$ 0 0	0 1 0
1 0 0	1 0 0	0 $\bar{1}$ 0	$\bar{1}$ 0 0	0 1 0	0 0 1
0 $\bar{1}$ 0	0 0 1	1 0 0	0 $\bar{1}$ 0	$\bar{1}$ 0 0	1 0 0
0 0 $\bar{1}$	0 $\bar{1}$ 0	0 0 1	0 0 1	0 0 1	0 1 0
0 1 0	0 0 $\bar{1}$	0 $\bar{1}$ 0	0 1 0	0 0 $\bar{1}$	0 0 1
0 0 1	$\bar{1}$ 0 0	0 0 1	0 0 $\bar{1}$	1 0 0	$\bar{1}$ 0 0
1 0 0	0 1 0	$\bar{1}$ 0 0	$\bar{1}$ 0 0	0 $\bar{1}$ 0	0 $\bar{1}$ 0
0 $\bar{1}$ 0	0 1 0	$\bar{1}$ 0 0	0 0 1	0 $\bar{1}$ 0	$\bar{1}$ 0 0
0 0 $\bar{1}$	1 0 0	0 0 1	0 $\bar{1}$ 0	$\bar{1}$ 0 0	0 0 $\bar{1}$
1 0 0	0 0 $\bar{1}$	0 1 0	1 0 0	0 0 $\bar{1}$	0 $\bar{1}$ 0

Table 2.2 gives an example of the operation by $\mathbf{T}_i$ on a misorientation matrix (see also Pumphrey and Bowkett, 1972; Randle and Ralph, 1975 for other examples). The elements of **M** do not change (because $\mathbf{T}_i$ consists only of 0's and 1's) but the rows of the matrix are permuted to new positions. From equations 2.8 and 2.9, and 2.25 the angle/axis pair and interface-planes can be computed for each degenerate matrix, and these are included in the Table. The 'starting' matrix gives an angle/axis of 50.13°/123 (which corresponds to a Σ = 39b CSL: see section 3.2). We see that it is not obvious from their diverse values that the 24 angle/axis pairs are essentially describing the same misorientation; however the multiplicity relationship *is* obvious from the fact that the 24 degenerate misorientation matrices are composed of the same elements.

The misorientation matrix gives only Θ/UVW and not the indices of the boundary plane. If the indices of N_1 are $1\bar{5}\bar{1}$, then from equation 2.12 we obtain $1\bar{1}\bar{1}$ for N_2. Turning now to the interface-plane notation for the 24 variants, we see in Table 2.2 that the 'starting' misorientation is $[1\bar{5}\bar{1}][1\bar{1}\bar{1}]$ 32.2°. When this is permuted by the symmetry operations, N_1 remains fixed (because grain 1 is the reference grain) while the *form* of N_2 and the twist angle φ change. Each of the 24 misorientation axes are shown on a stereogram in figure 2.8. Note that each *UVW* falls in a separate unit triangle, so that each triangle in a hemisphere of the projection contains one axis of

TABLE 2.2

THE 24 SYMMETRY-RELATED VARIANTS OF A SINGLE MISORIENTATION ($\Sigma = 39b$) IN THE FORM OF A MATRIX, ANGLE /AXIS AND INTERFACE-PLANE NOTATION

Matrix	$\theta°$ / *UVW*	N_1	N_2	$\varphi°$
.667 -.564 .487 .667 .744 -.051 -.333 .359 .872	50.1/123	$1\bar{5}\bar{1}$	$1\bar{1}\bar{1}$	32.2
.333 -.359 -.872 .667 .744 -.051 .667 -.564 .487	73.6/$\bar{1}\bar{3}2$	$1\bar{5}\bar{1}$	$1\bar{1}1$	49.6
-.667 .564 -.487 .667 .744 -.051 .333 -.359 -.872	153.8/$\bar{3}\bar{8}1$	$1\bar{5}\bar{1}$	$\bar{1}\bar{1}1$	147.8
-.333 .359 .872 .667 .744 -.051 -.667 .564 -.487	122.6/251	$1\bar{5}\bar{1}$	$\bar{1}\bar{1}\bar{1}$	114.0
-.667 -.744 .051 .667 -.564 .487 -.333 .359 .872	132.8/$\bar{1}$,3,11	$1\bar{5}\bar{1}$	$11\bar{1}$	92.2
-.667 .564 -.487 -.667 -.744 .051 -.333 .359 .872	140.3/$2\bar{1}\bar{8}$	$1\bar{5}\bar{1}$	$\bar{1}1\bar{1}$	87.8
.667 .744 -.051 -.667 .564 -.487 -.333 .359 .872	56.5/$31\bar{5}$	$1\bar{5}\bar{1}$	$\bar{1}\bar{1}\bar{1}$	6.0
.667 -.564 .487 .333 -.359 -.872 .667 .744 -.051	111.8/$9\bar{1}5$	$1\bar{5}\bar{1}$	$11\bar{1}$	27.8

Table 2.2 (*continued*)

.667	-.564	.487	167.0/$\bar{8}3\bar{2}$	$1\bar{5}\bar{1}$	111	152.2
-.667	-.744	.051				
.333	-.359	-.872				
.667	-.564	.487	87.8/$\bar{7}51$	$1\bar{5}\bar{1}$	$1\bar{1}1$	70.3
-.333	.359	.872				
-.667	-.744	.051				
-.333	.359	.872	167.0/546	$1\bar{5}\bar{1}$	$\bar{1}1\bar{1}$	152.2
.667	-.564	.487				
.667	.744	-.051				
.667	.744	-.051	75.1/$\bar{4}\bar{2}\bar{3}$	$1\bar{5}\bar{1}$	$\bar{1}\bar{1}1$	27.8
-.333	.359	.872				
.667	-.564	.487				
.333	-.359	-.872	126.2/$\bar{6}\bar{1}5$	$1\bar{5}\bar{1}$	111	32.2
.667	-.564	.487				
-.667	-.744	.051				
.667	.744	-.051	126.2/$73\bar{2}$	$1\bar{5}\bar{1}$	$\bar{1}1\bar{1}$	32.2
.333	-.359	-.872				
-.667	.564	-.487				
-.667	-.744	.051	153.8/$\bar{3}74$	$1\bar{5}\bar{1}$	$1\bar{1}\bar{1}$	152.2
-.333	.359	.872				
-.667	.564	-.487				
.333	-.359	-.872	94.4/$4\bar{5}\bar{1}$	$1\bar{5}\bar{1}$	$1\bar{1}\bar{1}$	87.8
-.667	.564	-.487				
.667	.744	-.051				
-.333	.359	.872	111.0/$\bar{1}6\bar{4}$	$1\bar{5}\bar{1}$	$\bar{1}\bar{1}1$	92.2
-.667	.564	-.487				
-.667	-.744	.051				
-.667	-.744	.051	140.3/$2\bar{4}7$	$1\bar{5}\bar{1}$	111	87.8
.333	-.359	-.872				
.667	-.564	.487				

Table 2.2 (*continued*)

-.667	.564	-.487	132.8/$\bar{1}\bar{9}\bar{7}$	$1\bar{5}\bar{1}$	$\bar{1}\bar{1}\bar{1}$	126.0
-.333	.359	.872				
.667	.744	-.051				
.667	.744	-.051	152.2/$\bar{1}\bar{1}$,$\bar{5}$,$\bar{1}$	$1\bar{5}\bar{1}$	$\bar{1}11$	87.8
.667	-.564	.487				
.333	-.359	-.872				
-.333	.359	.872	142.7/$\bar{3}1\bar{5}$	$1\bar{5}\bar{1}$	$\bar{1}11$	32.2
-.667	-.744	.051				
.667	-.564	.487				
-.667	.564	-.487	170.8/$57\bar{9}$	$1\bar{5}\bar{1}$	$\bar{1}11$	152.2
.333	-.359	-.872				
-.667	-.744	.051				
-.667	-.744	.051	170.8/5,$\bar{1}\bar{1}$,3	$1\bar{5}\bar{1}$	$1\bar{1}1$	169.6
-.667	.564	-.487				
.333	-.359	-.872				
.333	-.359	-.872	161.6/$5\bar{2}\bar{3}$	$1\bar{5}\bar{1}$	$11\bar{1}$	147.8
-.667	-.744	.051				
-.667	.564	-.487				

misorientation. This is the case for a general misorientation axis only, and not where *UVW* has higher symmetry. For convenience only the upper hemisphere has been used so that axes where *W* in $UV\bar{W}$ is negative have been rotated through 180°, e.g. $\bar{U}V\bar{W}$ has been converted to *UVW*.

In general it is usual to quote only one of the 24 equivalent values, particularly where only Θ/*UVW* and not *N* is measured. By convention the chosen variant is that which is associated with the smallest Θ; this value of Θ/*UVW* is known as the disorientation. As an alternative to using equation 2.20 to find the matrix variant which contains the disorientation, the rows of the matrix and the signs of elements within a particular row may be interchanged. From equation 2.8 it is clear that in order to find the matrix which corresponds to the disorientation it is necessary to change rows and signs of the matrix until its trace (i.e. $a_{11}+a_{22}+a_{33}$) is maximised.

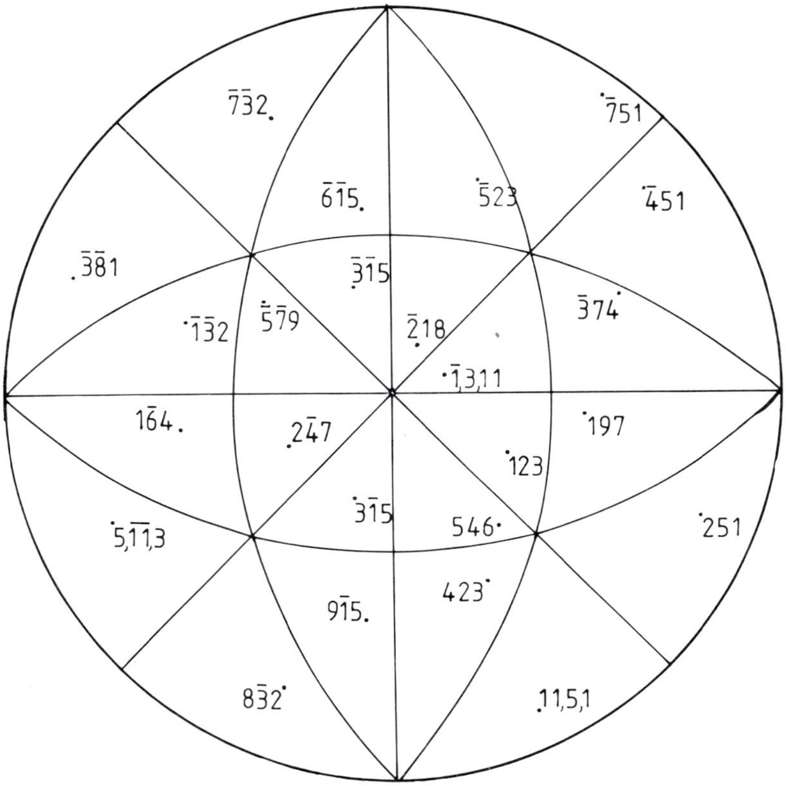

Figure 2.8 The misorientation axes for the 24 symmetry-related solutions of the angle/axis pair 50.13°/123 represented on a stereogram, showing that each of the 24 solutions falls in a different unit triangle. (50.13°/123 is a CSL having $\Sigma = 39b$).

2.5 TILT AND TWIST COMPONENTS

The total misorientation Θ/UVW can be decomposed into two sequential operations: a tilt rotation followed by a twist rotation. The tilt and twist components are characterised by *UVW* being parallel or perpendicular to the GB plane respectively (Lange, 1967; Wolf, 1985). The tilt angle component of the total misorientation is about an axis perpendicular to both N_1 and N_2 and the subsequent twist rotation is about N_2. This sequence is illustrated in figure 2.9. Figure 2.9a is a block of material containing a GB with plane normals N_1 and N_2, and reference axes for grain 1 $x_1y_1z_1$. A tilt rotation of ψ about an axis perpendicular to both N_1 and N_2, n_T, aligns the two normals. n_T and ψ are given by

$$n_T = (N_1 \times N_2)/ \mid (N_1 \times N_2) \mid \quad (2.22)$$

$$\sin \psi = \mid (N_1 \times N_2) \mid \quad (2.23)$$

Figure 2.9b shows the the positions of the coordinate axes of both grains reoriented after the tilt so that N_1 and N_2 are now parallel. Finally in figure 2.9c a twist rotation of φ is performed about the common direction N_1 and N_2 (as already illustrated in figure 2.1b) to rotate $x_1y_1z_1$ onto $x_2y_2z_2$. Hence the total misorientation may be written as

$$\mathbf{M}(\Theta, UVW) = \mathbf{M}(N_2,\varphi)\, \mathbf{M}(n_T,\psi) \quad (2.24)$$

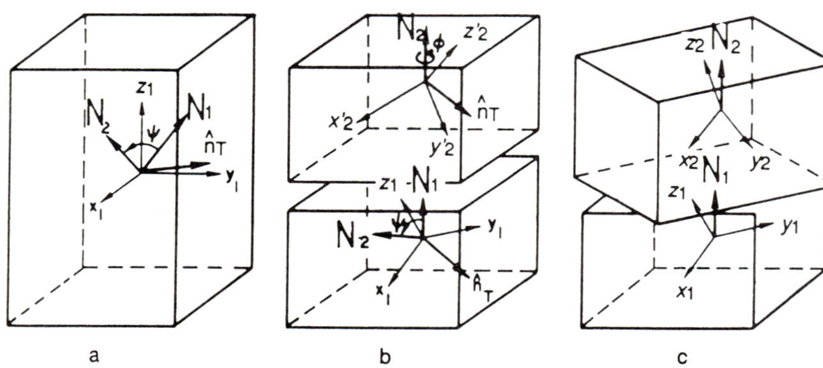

Figure 2.9 Sequence showing the specification of a general GB as a tilt rotation followed by a twist rotation. (a) shows a block of material with axes $x_1y_1z_1$ containing a GB with plane normals N_1 and N_2 in each of the grains. A rotation through ψ about the tilt axis n_T aligns N_1 and N_2 parallel, as shown in which is shown in (c). (Wolf and Lutsko, 1989).

A simple way to illustrate the relationship in equation 2.24 is by use of spherical trigonometry. $\mathbf{M}(\Theta, UVW)$, $\mathbf{M}(N_2,\varphi)$ and $\mathbf{M}(n_T,\psi)$ can be represented by spherical triangles having apices (N_1,N_2,UVW), (N_1,UVW,n_T) and (N_1,N_2,n_T) respectively as shown in figure 2.10. The algebra is simplified considerably by noticing that, by definition, the tilt and twist axes are perpendicular to each other, i.e. N_1 and N_2 are 90° from n_T. Hence we obtain two identical Napierian (i.e. right-angled spherical) triangles, with apices n_T,N_1,n' and n_T,N_2,n' where n' is the right-angle when the tilt angle ψ is bisected (see Appendix and McKie and Mckie (1974) for a guide to spherical trigonometry). Hence from consideration of the relevant triangles the tilt and twist angles are obtained from

$$\sin(\psi/2) = \sin(\Theta/2)\ \sin|(N_1 x UVW)| \qquad (2.25a)$$

$$\sin(\varphi/2) = \frac{\sin(\psi/2)\ \sin|(n_T x UVW)|}{\sin|(N_1 x UVW)|} \qquad (2.25b)$$

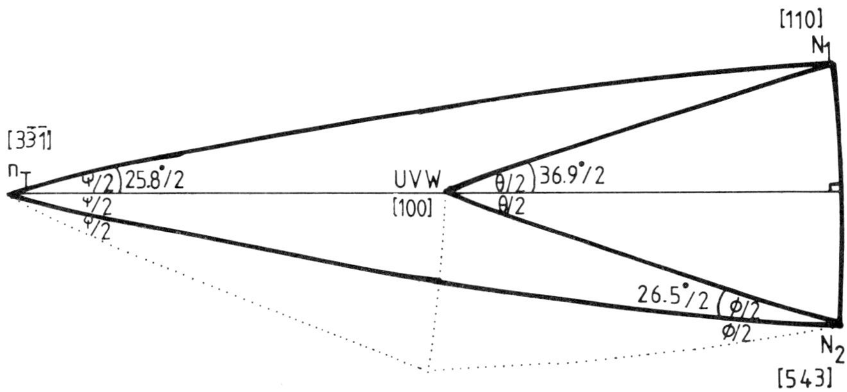

Figure 2.10 Illustration of the decomposition of a misorientation into tilt and twist components using spherical trigonometry. The calculations are simplified by bisecting Θ, φ and ψ to give several right-angled triangles. The values of the misorientation, tilt and twist parameters are labelled for this particular example, which is referred to again in section 3.2.5.

The relationship between Θ, φ, and ψ is given by

$$1 + \cos\varphi = 2(1 + \cos\Theta)/(1 + \cos\psi) \qquad (2.26)$$

Some GBs do not decompose into tilt and twist components because either φ or ψ is zero. The boundary is then either pure tilt or pure twist as follows (Wolf and Lutsko, 1989):

$$\psi = 0, \text{ then } UVW = N_I \text{ and } \Theta = \varphi \text{ (pure twist)} \tag{2.27a}$$

$$\varphi = 0, \text{ then } UVW = n_T \text{ and } \Theta = \psi \text{ (pure tilt)} \tag{2.27b}$$

Tilt and twist GBs are discussed in more detail in section 3.2.5.

3

THEORETICAL ASPECTS OF GRAIN BOUNDARY GEOMETRY II: PERIODIC GEOMETRIES

3.1 INTRODUCTION

In chapter 2, GB geometry was described for the general case, that is, GBs with a non-specific angle/axis pair and plane normal. We now introduce a subset of GBs whose geometry is not random because 'matching' in the interfacing lattices introduces a periodicity. Such GBs are sometimes referred to as 'geometrically special' (Priester, 1989) and are important because of their potential association with 'special' GB properties, as was outlined in chapter 1.

From a practical viewpoint, the most convenient and widely accepted means of recognition of geometrically special GBs in polycrystals is the coincidence site lattice (CSL) formalism, and therefore most of this chapter is devoted to a detailed explanation of this model and its role in the categorisation of GB geometry. For completeness, an overview of some other models which relate to GB geometry and/or structure is included in section 3.3.

3.2 THE COINCIDENCE SITE LATTICE

To understand how non-random, periodic GB geometries arise it is instructive to consider two interpenetrating lattices with a common origin point. This concept was introduced in section 2.2.2. Certain specific combinations of Θ and *UVW* result in the coincidence of a proportion of lattice points from each lattice. Because an array of lattice points is necessarily periodic in three-dimensions, the superlattice of points which coincide as a result of specific misorientations is also periodic. This is illustrated in figure 3.1a which shows a projection of an interpenetrating set of points contributed by two lattices - called a dichromatic pattern (Pond and Bollmann, 1979) - distinguished as open circles (lattice 1) and circles containing dots (lattice 2). The lattice are misoriented by 36.87°/001, and 001 is perpendicular to the plane of the paper. This specific misorientation gives rise to an array of lattice points which coincide - a coincidence site lattice, CSL, - shown as filled circles in figure 3.1a. In figure 3.1b the generation of a GB from these two interpenetrating lattices, by removing symbols from each side of the GB as in figure 2.4b, is shown. Finally in figure 3.1c the coincidence sites are joined to emphasise their periodicity. The ratio of the volume of the CSL unit cell to the volume of the lattice cell is 5 in this case. In other words 1 in 5 lattice sites are in coincidence.

If we change the angle of misorientation about the same 001 axis to 22.62°, a new superlattice of coincidence sites is obtained, as shown in figure 3.1d. For this new Θ/UVW 1 in 13 lattice sites are in coincidence, i.e. the volume ratio is 13. This volume ratio is given the symbol Σ. Other Σ-values are generated from different, specific combinations of Θ and *UVW*. Figure 2.4, which shows two lattices misoriented by 38.94°/0$\bar{1}$1, has 1 in 9 coincidence sites which is referred to as a 'Σ = 9 CSL'; the stereogram in figure 2.3 shows Θ/UVW = 60.0°/111 which is a Σ = 3 CSL. Table 3.1 shows a list of specific Θ/UVWs which generate CSLs up to Σ = 49. Note that only odd values of Σ arise for cubic crystals (Grimmer *et al*, 1974).

TABLE 3.1

ANGLE AND AXIS OF MISORIENTATION, TWINNING PLANES AND MAXIMUM DEVIATION (BRANDON CRITERION) FOR CSLs UP TO Σ = 49

Σ	Θ° / *UVW*	Twin planes		v_m
3	60/111	111	211	8.67
5	36.87/100	210	310	6.71
7	38.21/111	321		5.67
9	38.94/110	221	411	5.00
11	50.48/110	311	332	4.52
13a	22.62/100	320	510	4.16
13b	27.80/111	431		4.16
15	48.19/210	521		3.87
17a	28.07/100	410	530	3.64
17b	61.93/221	322	433	3.64
19a	26.53/110	331	611	3.44
19b	46.83/111	532		3.44
21a	21.79/111	541		3.27
21b	44.40/211	421		3.27

Table 3.1 (*continued*)

23	40.45/311	631		3.13
25a	16.25/100	430	710	3.00
25b	51.68/331	543		3.00
27a	31.58/110	511	552	2.89
27b	35.42/210	721		2.89
29a	43.61/100	520	730	2.79
29b	46.39/221	432		2.79
31a	17.90/111	651		2.69
31b	52.19/211	732		2.69
33a	20.05/110	441	811	2.61
33b	33.55/311	741		2.61
33c	58.98/110	522	554	2.61
35a	34.04/211	531		2.50
35b	43.23/331	653		2.50
37a	18.92/100	610	750	2.47
37b	43.13/310	831		2.47
37c	50.57/111	743		2.47
39a	32.21/111	752		2.40
39b	50.13/321	-		2.40
41a	12.68/100	540	910	2.34
41b	40.88/210	621		2.34
41c	55.88/110	443	833	2.34
43a	15.18/111	761		2.29
43b	27.91/210	921		2.29
43c	60.77/332	533	655	2.29
45a	28.62/311	851		2.24
45b	36.87/221	542		2.24
45c	53.13/221	754		2.24
47a	37.07/331	763		2.19
47b	43.66/320	932		2.19
49a	43.58/111	853		2.14
49b	43.58/511	941		2.14
49c	49.22/322	632		2.14

Although this book is concerned only with GB geometry in cubic polycrystals, it should be mentioned that CSLs can also occur in non-cubic materials. There are several publications which give details of the extention of the CSL model to other crystal systems, particularly hexagonal (Bleris *et al,* 1982; Grimmer, 1989; Grimmer *et al,* 1990). Furthermore, there is now interest in extending the CSL model to the important high temperature superconductors, whose crystal structure is orthorhombic, because GB

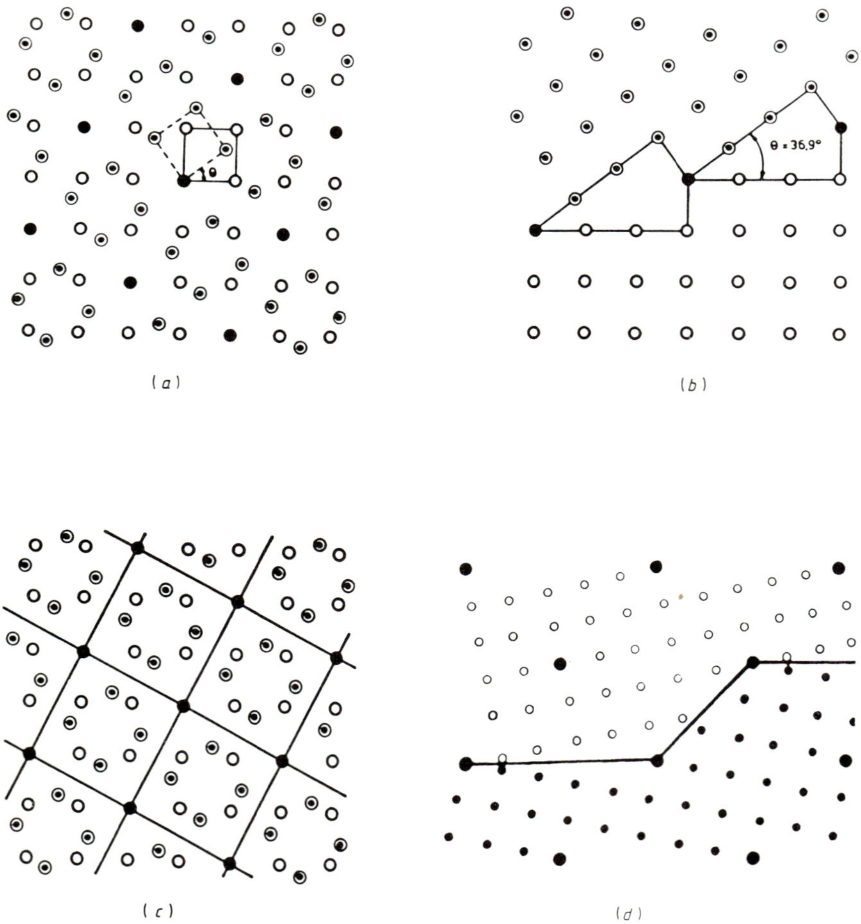

Figure 3.1 (a) Two interpenetrating lattices, misoriented by 36.87°/001 forming a dichromatic pattern, viewed down 001. (b) GB generated from the dichromatic pattern in (a). (c) coincidence site lattice, $\Sigma = 5$, generated by the misorientation shown in (a) and (b). (Fischmeister, 1985). (d) GB formed by a misorientation of 22.62°/001, which gives a CSL of $\Sigma = 13$. The coincidence sites are denoted by solid symbols throughout.

structure in these materials is critical to development of the optimal superconducting properties (Zhu *et al,* 1991). The determination of CSLs in crystal systems other than cubic is more complex because it depends critically on the lattice parameter axial ratios. The concept of 'near coincidence' (Bonnet *et al,* 1981) or 'constrained coincidence' (Singh *et al,* 1990) has been developed to extend the range of allowable CSLs.

3.2.1 Physical significance of the coincidence site lattice

The significance of the CSL is that the periodicity which it introduces in two notionally interpenetrating lattices becomes a physical reality at the GB itself, and in turn the atomic arrangements in the GB are non-random. The geometry is not, of course, determined solely by the misorientation; the inclination of the boundary plane (see section 2.2.2) controls how many coincident sites occur *actually in the GB plane itself,* i.e. where the GB is 'placed' in the interpenetrating lattices (Paidar, 1987). This is discussed further in sections 3.2.4 and 3.2.5. A random plane section through the CSL will have a density of 1 in Σ coincidence sites.

A direct consequence of the periodicity associated with a CSL relationship at a GB is that, put simply, the two meeting interfaces can sometimes fit together more closely than would be the case for random geometry. This 'good fit' property is enhanced by atomic relaxations at the GB, mostly brought about by rigid body translations (Fischmeister, 1985). Such relaxations adjust the positions of coinciding atoms in and near the boundary; however the periodicity (= Σ), which is the significant parameter, is maintained. Figure 3.2 is an atomic resolution micrograph which shows a $\Sigma = 5$ CSL GB and illustrates the effect of translations in the GB plane on its core structure (Merkle, 1989; Merkle and Smith, 1987).

CSL boundaries were first reported in connection with the misorientations observed during secondary recrystallisation in copper (Kronberg and Wilson, 1949). Later the model was extended and formalised (Brandon *et al,* 1964; Brandon, 1966), and it was suggested that the 'CSL model' could explain the observed anisotropy in structure-sensitive GB properties such as energy, segregation, diffusion and migration. For some years this approach, with modifications to incorporate such factors as deviations from exact coincidence at GBs (see section 3.2.3), and inclination of the GB plane (section 3.2.4 and 3.2.5), yielded some success with respect to linking GB properties to geometry. There are several reviews which summarise aspects of the progress in knowledge of GB property/geometry relationships made from these investigations. (Chadwick and Smith, 1976; Goodhew, 1980; Ralph, 1980; Watanabe, 1984; Fischmeister, 1985; Sutton and Balluffi, 1987; Priester, 1989). However, the CSL description of GB geometry does not predict absolutely GB properties for the complete range of materials and conditions (Sutton and Balluffi, 1987). The reason for this is that GB properties depend

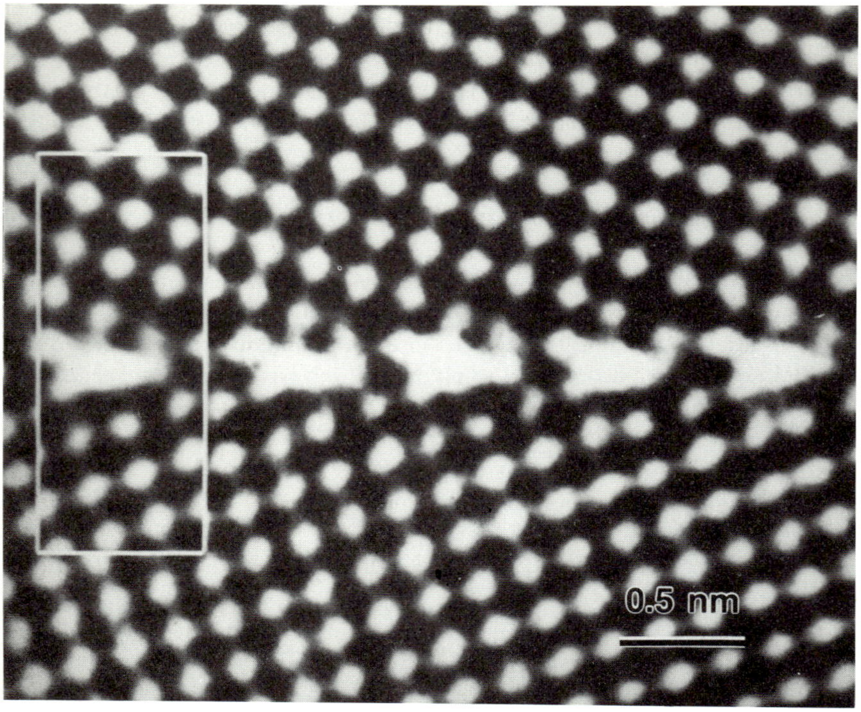

Figure 3.2 Atomic resolution micrographs of a symmetrical tilt (310) GB in the $\Sigma = 5$ system. (Courtesy of K.L. Merkle).

only partially on GB geometry; they depend also upon local defect structure, chemistry and electronic bonding.

The CSL model is widely used in connection with GB geometry, although from the foregoing paragraph it is clear that the geometrical component of GB structure, especially if only the grain misorientation is measured, is not sufficient information to predict completely 'special' properties. Where the orientation of the GB plane is measured in addition to the misorientation, the CSL model is more powerful in its application to prediction of GB properties because the geometrical information is then directly applicable to the GB itself (Paidar, 1987; Wolf, 1985; Randle, 1989). Notwithstanding the limitations of the CSL approach, it is a convenient and accessible method for the categorisation of GB geometry (see Table 7.1, chapter 7) and so has remained one of the cornerstones of GB studies. Certainly the knowledge that a GB is near a low-Σ CSL is an indicator that the GB has the potential for 'special' characteristics (Varin, 1979; Goodhew and Smith, 1980).

There is no consensus concerning an upper limit for Σ which is significant in terms of when the boundary periodicity becomes too large to confer 'potential specialness' on the boundary (Shvindlerman and Straumal, 1985; Dechamps *et al,* 1987). In the literature there are examples of CSL classification limits extending from Σ = 19 upwards. These are listed in Table 7.1 and discussed in section 7.3.

3.2.2 Mathematical analysis of CSL geometry

A generating function allows us to calculate all values of Σ and accompanying angles of misorientation, Θ, for specific axes of misorientation, *UVW*, for cubic crystals of any lattice type, i.e. both face-centered cubic (fcc) and body-centered cubic (bcc) (Ranganathan, 1966):

$$\Sigma = x^2 + N y^2 \qquad (3.1a)$$

$$\tan(\Theta/2) = y N^{1/2}/x \qquad (3.1b)$$

where $N = U^2 + V^2 + W^2$ and x and y are integers ≥ 0. If the Σ-value obtained from equation 3.1a is an even number it must be divided by 2 iteratively until an odd number results. Table 3.2 illustrates the use of the generating function for some values of x and y on the 321 axis. There is no special significance attached to this axis; it has been selected for demonstration purposes only.

In practice, the only CSLs which are likely to be of interest in either fcc or bcc materials are those with relatively low Σ-values. The Θ/*UVW*s associated with these low-Σ CSLs are tabulated in several publications (Pumphrey and Bowkett, 1971; Acton and Bevis, 1971; Grimmer *et al,* 1974; Mykura, 1980; Randle *et al,* 1988; Forwood and Clarebrough, 1992) and also in this book. The generating function therefore rarely needs to be used. Table 3.3 lists Θ/*UVW*s for CSLs with Σ up to 31. As discussed in section 2.4, a misorientation matrix (and hence Θ/*UVW*) has 24 equivalent forms: Table 3.3 includes all 24 forms for each Σ value up to 31, listed according to *UVW*. In other references, all 24 forms for CSLs are listed according to Σ-value up to 43 (Mykura, 1980; Forwood and Clarebrough, 1992). Mykura (1980) also lists the disorientation Θ_D/*UVW*, the twinning planes and rotation quaternions (see below) for Σ up to 101. Table 3.4 contains the disorientation, listed according to *UVW*, for CSLs with Σ up to 49. From a knowledge of the disorientation the other 24 symmetry-related variants can be generated using equations 2.8 to 2.11.

TABLE 3.2

RANGANATHAN'S GENERATING FUNCTION FOR *UVW* = 123

Θ°	Σ	*x*	*y*
180	7	0	1
0	1	1	0
150.07	15	1	1
123.75	9	2	1
102.56	23	3	1
86.18	15	4	1
164.78	57	1	2
150.07	15	2	2
136.31	65	3	2
123.75	9	4	2
169.82	127	1	3
159.79	65	2	3
150.07	135	3	3
140.77	71	4	3
172.35	225	1	4
164.78	57	2	4
157.33	233	3	4
150.07	15	4	4

TABLE 3.3

All values of Θ/*UVW* for Σ up to 31

	Σ	Θ		Σ	Θ		Σ	Θ
100	5	36.9	110	3	70.5	111	3	60
	13a	22.6		9	38.9		7	38.2
	17a	28.1		11	50.5		13b	27.8
	25a	16.3		17b	86.6		19b	46.6
	29b	43.6		19a	26.5		21a	21.8
				27b	31.6		31a	17.9

Table 3.3 (*continued*)

210	3	131.8	211	3	180	221	5	143.1
	5	180		5	101.5		9	90
	7	73.4		7	135.6		9	180
	9	96.4		11	63.0		13b	112.2
	15	48.2		15	78.5		17b	61.9
	21b	58.4		21b	44.4		25b	73.7
	23	163.0		25b	156.9		29a	46.4
	27a	35.4		29a	149.6			
	29a	112.3		31b	52.2			
310	5	180	311	3	146.4	320	7	149.0
	7	115.4		5	95.7		11	100.5
	11	144.9		9	67.1		13a	180
	13b	76.7		11	180		17b	122.0
	19a	93.0		15	50.7		19b	71.6
	23	55.6		15	117.8		29a	84.1
				23	40.5		31b	54.5
				25b	168.3			
				27a	79.3			
				31b	126.6			
321	7	180	322	9	152.7	410	9	152.7
	9	123.8		13a	107.9		13b	107.9
	15	86.2		17b	180		17a	180
	15	150.1		21a	128.3		21a	79.0
	23	102.6		21b	79.0		21b	128.3
	25b	63.9						
411	9	180	331	5	154.2	421	11	155.4
	11	129.5		7	110.9		15	113.6
	17a	93.4		11	82.2		21b	180
	19b	153.5		17b	63.8		23	85.0
	27a	109.5		19a	180		25b	132.8
	27b	70.5		23	130.7			
				25b	51.7			
332	11	180	430	13b	157.4	431	13b	180
	13a	133.8		17b	118.1		15	137.2
	19a	99.1		25a	180		21b	103.8
	23	155.9		25b	90		27a	157.8
	29a	76.0		29a	136.4		31b	80.7
	31a	114.8						

Table 3.3 (*continued*)

510	13a	180	511	7	158.2	432	15	159.0
	15	137.2		9	120.0		19a	121.8
	21a	103.8		13a	92.2		27a	94.3
	27b	157.8		19a	73.2		29a	180
	31a	80.7		27a	60			
				27b	180			
				31b	137.9			
520	15	159.0	521	15	180	441	17a	160.3
	19b	121.8		17b	139.9		21b	124.9
	27b	94.3		23	107.7		29a	97.9
	29b	180		31b	159.3			
522	17b	160.3	433	17b	180	530	17a	180
	21b	124.9		19a	142.1		19b	142.1
	29b	97.9		25a	111.1		25b	111.1
610	19a	161.3	532	19b	180	611	19a	180
	23	127.5		21b	144.1		21b	144.1
	31a	101.2		27a	114.0		27b	114.0
443	21b	162.3	540	21a	162.3	621	21b	162.3
	25a	129.8		25b	129.8		25b	129.8
531	9	160.8	533	11	162.7	551	13a	164.1
	11	126.2		13b	130.8		15	134.4
	15	99.6		17a	105.3		19b	110.0
	21b	80.4		23	86.3		25b	91.2
	29a	66.6		31b	72.2			
541	21a	180	542	23	163	631	23	180
	23	145.7		27a	131.8		25b	147.1
	29a	116.6					31b	118.9
632	25b	163.7	543	25b	180	710	25a	180
	29a	133.6		27a	148.4		27a	148.4
711	13b	164.1	553	15	165.2	731	15	165.2
	15	134.4		17a	137.3		17b	137.3
	19a	110.0		21a	113.9		21b	113.9
	25a	91.2		27b	95.3		27a	95.3

Table 3.3 (*continued*)

641	27b	164.4	720	27a	164.4	552	27b	180
	31b	135.2		31b	135.2		29b	149.6
721	27a	180	544	29a	164.9	730	29b	180
	29a	149.6	722	29a	164.9		31b	150.6
733	17b	166.1	751	19a	166.8	753	21b	167.5
	19b	139.7		21b	141.8		23	143.6
	23	117.2		25b	120		27b	122.5
	29b	98.9		31b	102.1			
911	21a	167.5	931	23	168.0	755	25b	168.5
	23	143.6		25b	145.1		27b	146.4
	27a	122.5		29a	124.7		31a	126.6
771	25a	168.5	773	27a	169.0	951	27a	169.0
	27a	146.4		29b	147.7		29a	147.7
	31b	126.6	645	31b	165.4	650	31a	165.4
953	29a	169.4	775	31b	169.7	11,1,1	31a	169.7
	31b	148.7	732	31b	180	651	31a	180

The significance of the letters which occur in Tables 3.1, 3.3 and 3.4 are that more than one misorientation can generate geometrically independent CSLs with a particular Σ-value. For example a $\Sigma = 39$ CSL arises from the disorientation 32.21°/111 ($\Sigma = 39$a) and also 50.13°/123 ($\Sigma = 39$b). A full list of all 24 variants of $\Sigma = 39$b were given in Table 2.1. The letters which distinguish identical Σ-values are designated according to increasing Θ.

In analogy to equation 2.6 the misorientation matrix for a CSL is given by

$$\mathbf{M}_{CSL} = 1/\Sigma \begin{bmatrix} A_{11} & A_{12} & A_{13} \\ A_{21} & A_{22} & A_{23} \\ A_{31} & A_{32} & A_{33} \end{bmatrix} \tag{3.2}$$

Hence the misorientation matrix in equation 2.7, which refers to a $\Sigma = 3$ CSL, could be rewritten:

$$\mathbf{M} = 1/3 \begin{bmatrix} 2 & -1 & 2 \\ 2 & 2 & -1 \\ -1 & 2 & 2 \end{bmatrix} \tag{3.3}$$

where the matrix elements are all integers and have no common factors with Σ. This matrix is shown on a stereogram in figure 2.3.

There are many mathematical subtleties associated with CSLs. For instance, Σ can be expressed in terms of four numbers, A,B,C,D, sometimes called a rotation quaternion (Mykura, 1980; Morawiec and Pospiech 1989), such that:

$$A^2 + B^2 + C^2 + D^2 = \Sigma \tag{3.4}$$

The disorientation Θ_D/UVW is then given by one of the following equations:

$$\Theta \text{ and } U,V,W = 2\cos^{-1}(A/\Sigma^{-1/2}) \text{ and } B,C,D \tag{3.5a}$$

$$\Theta \text{ and } U,V,W = 2\cos^{-1}((A+B)/(2\Sigma)^{1/2}) \text{ and } (A-B),(C+D),(C-D) \tag{3.5b}$$

$$\Theta \text{ and } U,V,W = 2\cos^{-1}((A+B+C+D)/2\Sigma^{1/2}) \text{ and } (A+B-C-D),(A-B+C-D),(A-B-C+D) \tag{3.5c}$$

For example the quaternion for $\Sigma = 15$ is 3,2,1,1. Solutions to equations 3.5a, b and c are 78.46°/211, 48.19°/120 and 117.82°/311 respectively. In this case the second solution gives the disorientation.

For some CSLs the highest angle of misorientation in the set of 24 is 180°: this corresponds to $x = 0$ $y = 1$ in equation 3.1. The misorientation axis associated with $\Theta = 180°$ is known as the twin axis. When this axis lies parallel to the GB plane a rotation through 180° about the twin axis is equivalent to a reflection across the GB plane, which is then known as the twin plane (Christian, 1975). The twin plane, or one of the twin planes if more than one exists for a particular CSL, is a plane in which every site is a coincidence site. In other words the twin plane is a close-packed plane in the CSL. The role of the GB plane in the CSL is discussed in detail in sections 2.3.4 and 3.2.5. The twin axes for CSLs up to $\Sigma = 49$ are given in Table 3.1. The lowest value of Σ for which there is no twinning plane (because there is no solution to equation 3.1a) is 39. The indices of $(UVW)_T$ when $\Theta = 180°$, i.e. the indices of the twin axis or twin plane, are given by

$$U^2 + V^2 + W^2 = \Sigma \text{ or } 2\Sigma \qquad (3.6)$$

The most well known example of a twin rotation is the $\Sigma = 3$ case. In practice, only this CSL is usually referred to as a 'twin'. The twin planes are 111 and 112, and from equation 3.6 $\Sigma = 1^2+1^2+1^2$ or $2\Sigma = 1^2+1^2+2^2$.

TABLE 3.4

VALUES OF θ°/*UVW* FOR CSLS UP TO Σ = 49 GROUPED ACCORDING TO *UVW*

100		110		111	
θ°	Σ	θ°	Σ	θ°	Σ
12.7	41a	20.1	33a	15.2	43a
16.3	25a	26.5	19a	17.9	31a
19.0	37a	31.6	27a	21.8	21a
22.6	13a	38.9	9	27.8	13b
28.1	17a	50.5	11	32.2	39a
36.9	5	55.9	41c	38.2	7
46.3	29a	59.0	33c	46.8	19b
				50.6	37c
				60.0	3
210		**221**		**221**	
27.9	43b	34.0	35a	36.9	45b
35.4	27b	44.4	21b	46.4	29b
40.9	41b	52.2	31b	53.1	45c
48.2	15			61.9	17b
310		**311**		**331**	
43.1	37b	28.6	45a	43.2	35b
		33.6	33b	51.7	25b
		40.5	23		
332		**321**			
60.8	43c	50.1	39b		

3.2.3 Deviations from exact coincidence

In polycrystals there is little chance of a GB misorientation being exactly that of a CSL. However, the 'special' properties which may be associated with low-Σ CSLs are observed to occur in GBs which are close to, but not exactly, CSL misorientations. Hence the requirement for the analysis of GB geometry in polycrystals is to identify all GB misorientations which are 'close to' CSLs in terms of an angular deviation. There is no definitive criterion which stipulates a maximum for this angular deviation, although there are physically-based guidelines which have been used to propose values of the maximum deviation from a CSL, v_m. We will examine briefly the factors which affect v_m.

The important feature of GBs which are close to CSLs is that, for small angular deviations, the CSL is conserved by dislocation arrays known as secondary intrinsic GB dislocations (Bollmann, 1970). The existence of these dislocations has been observed and studied in a wide range of materials, and figure 3.3a shows an example from a $\Sigma = 3$ GB in a superalloy. The existence of secondary intrinsic dislocations in high angle GBs is analogous to that of primary intrinsic GB dislocations in low angle GBs (figure 3.3b). The dislocation description of low angle GBs is well established (Read and Shockley, 1950). Low angle GBs consist of dislocation arrays which are able to compensate entirely for the small angular difference - usually taken to be up to 10-15° (Dechamps *et al,* 1987) - between grains. The analysis of GB dislocations is discussed in detail elsewhere (Forwood and Clarebrough, 1992).

The structural similarities between CSL GBs and low angle GBs led to the adoption for the CSL case of the following well-known relationship for the low angle case (i.e. the low angle approximation of the Read-Shockley relationship (Read and Shockley, 1950):

$$\delta = \boldsymbol{b}/d \text{ (low angle)} \tag{3.7a}$$

$$v_m = \boldsymbol{b}/d \text{ (CSL)} \tag{3.7b}$$

where d is the dislocation density in the GB, $\boldsymbol{b}$ is the Burger's vector of the dislocations and δ is the angular misorientation of the low angle GB. Thus, v_m corresponds to the highest density (i.e. smallest spacing) of dislocations possible in the GB. The density of dislocations which can be accommodated in the GB is related to its periodicity (i.e. Σ) (Pumphrey, 1976); the smaller the period, the greater the density of dislocations which can 'fit' into the boundary in a regular, intrinsic array. The variation of v_m with Σ is usually taken to be as $\Sigma^{-1/2}$, which corresponds to a relationship based on the periodicity alone (Brandon, 1966). Hence

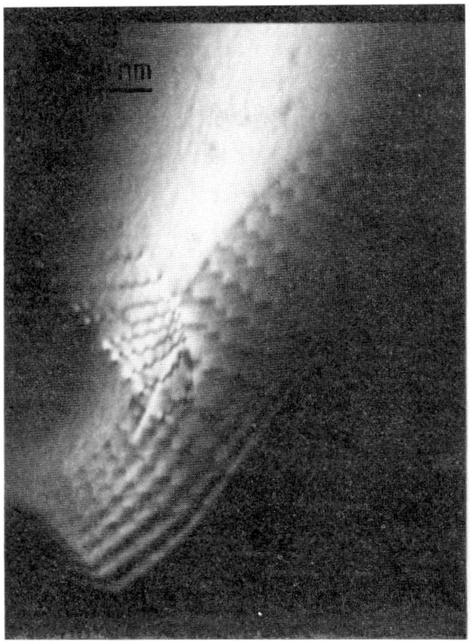

(a)

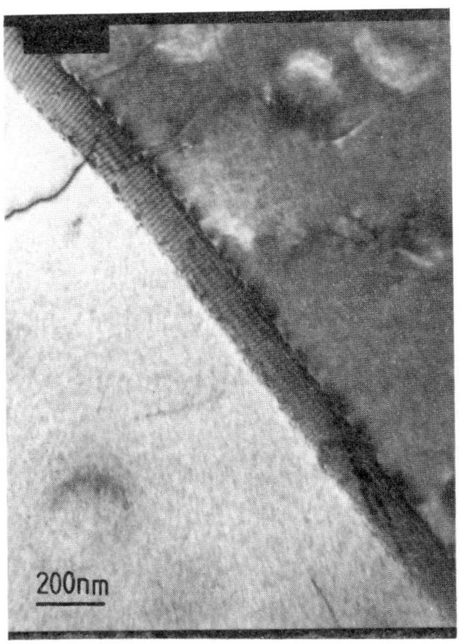

(b)

Figure 3.3 GBs with dislocation substructures (Randle and Ralph, 1988b). (a) A $\Sigma = 3$ GB with secondary intrinsic GB dislocations. There is a coherent precipitate in the GB. (b) Primary intrinsic GB dislocations in a low angle GB (Randle and Ralph, 1988b).

$$v_m = v_0 \Sigma^{-1/2} \tag{3.8}$$

where v_0 is a proportionality constant equal to the angular limit for a low angle boundary, 15°. Hence if $\Sigma = 1$ (meaning total coincidence between two lattices or a small deviation from it, i.e. a low angle GB) is substituted in equation 3.8 v_m is 15°, which is the limit on the misorientation angle of a low angle boundary. Table 3.1 includes values of v_m for Σ up to 49.

Equation 3.8 is known as the 'Brandon criterion' and has become established as the almost universally adopted method for deciding v_m (see Chapter 7). Using this criterion, for example, v_m for a $\Sigma = 3$ boundary is 8.66°, although 'annealing twin' GBs would normally be expected to be close to exact CSL matching (Randle, 1991a). Another criterion proposes that v_m should be proportional to $\Sigma^{-2/3}$ (Pumphrey, 1976) and another to Σ^{-1} for twist GBs (Ishida and McLean, 1973). There is evidence for the existence of secondary GB dislocations in GBs with greater angular deviations than v_m according to the Brandon criterion (Balluffi and Tan, 1972). However, one instance of observed correlation of special properties on deviation from a CSL showed a $\Sigma^{-5/6}$ dependence (Palumbo and Aust, 1990a). This dependence is also supported by a geometric argument: d is proportional to the mean edge of the CSL cell, $\Sigma^{1/3}$, and $\boldsymbol{b}$ is considered to be proportional to $\Sigma^{-1/2}$. Substitution of these values for $\boldsymbol{b}$ and d in equation 3.7 gives $\Sigma^{-5/6}$. The question of a definitive theoretically derived value for v_m remains unresolved, although there is evidence from TEM observations of intrinsic GB dislocations that the Brandon criterion may be too permissive and a $\Sigma^{-5/6}$ criterion is consistent with many experimental observations, particularly for $\Sigma < 27$ (figure 3.4). The criteria for the determination of the 'geometrical specialness' of GBs, i.e. the maximum allowable deviation from a CSL, are discussed in more detail elsewhere (Dechamps *et al,* 1987; Aust and Palumbo, 1991; Palumbo and Aust, 1992).

3.2.4 Planar coincidence site density

The density of coincidence sites in the GB itself, known as the planar coincidence site density, PCSD, depends critically upon the inclination of the GB with respect to the CSL (Brandon *et al,* 1964; Bishop and Chalmers, 1968; Christian, 1975); Goodhew *et al,* 1978). The PCSD is a rational factor of Σ, and is unity when all sites in the GB plane are coincidence sites. This case is the twinning plane which was introduced in section 3.2.2. At the other extreme, for every plane where PCSD = 1 there are (Σ-1) planes which do not contain any coincidence sites, i.e. PCSD = 0. Taking $\Sigma = 15$ as an example, the density of coincidence sites at the GB ranges from 1 through 1/3, 1/5, 1/15 to 0, and a randomly selected plane through the CSL will have a density of coincidence sites of 1 in Σ, as pointed out in subsection 3.2.1. In the same way

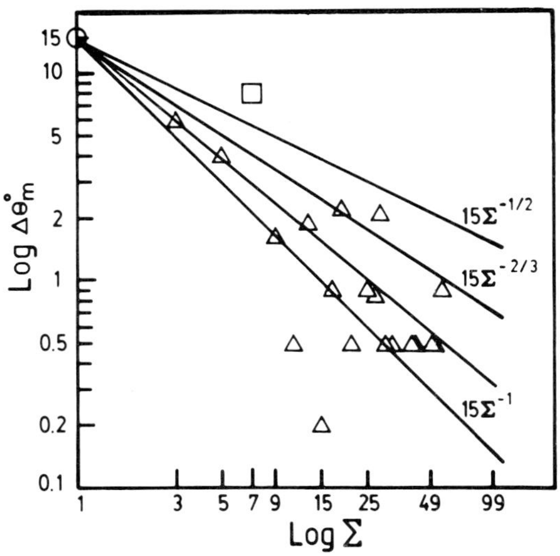

Figure 3.4 Maximum deviation angle from CSLs at which discrete intrinsic GB dislocations have been identified (Palumbo and Aust, 1992). A line denoting $15\Sigma^{-5/6}$ has been added.

that Σ is expressed as the *reciprocal* density of coincidence sites, the PCSD is also often expressed in the same way, i.e. 1, 3, 5, 15, or infinity for the example here. A list of planes having a high density of coincidence sites for Σ up to 31 can be found elsewhere (Acton and Bevis, 1971).

Clearly the position of the GB with respect to each interfacing lattice can influence greatly the 'good fit' of the GB and hence can have a marked effect on GB properties. The magnitude of the geometrical effect for various Σ-values can be seen by expressing the PCSD as the number of coincidence sites per unit area of GB, which is usually denoted by the symbol σ:

$$\sigma = B/(h^2+k^2+l^2)^{-1/2} \tag{3.9}$$

where B is the number of atoms per unit cell in the lattice (i.e. 4 for fcc and 2 for bcc) and *hkl* are the indices of the GB plane. Hence σ is affected by the lattice type although Σ is not. The most frequently quoted example is that for the twinning plane (111) in the Σ = 3 CSL for fcc crystals. Here $\sigma = 2.31a^2$, where a is the lattice parameter. For Σ >3 σ drops sharply, with the next highest value being the (311) twinning plane in Σ = 11, where $\sigma = 1.21a^2$. We recall that the highest σ is always on the twinning plane because here every lattice site is a coincidence site. The reciprocal of σ is the minimum area per coincidence site in the GB. Figure 3.5 shows the variation of $1/\sigma$ with Σ for twin planes in CSLs, fcc and bcc, for Σ up to 49.

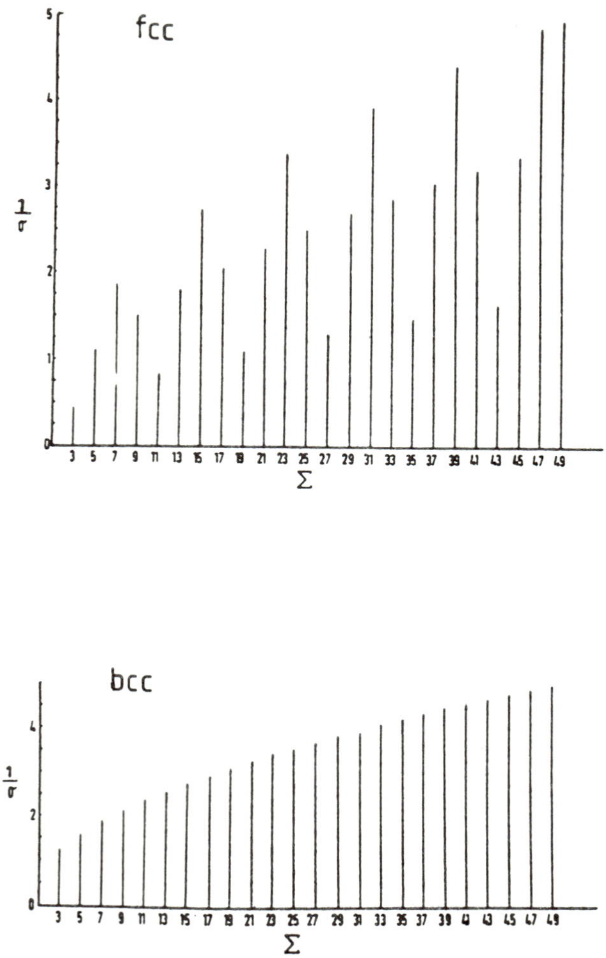

Figure 3.5 Variation of the minimum area in the twin plane per coincidence site ($1/\sigma$) in units of a^2 versus Σ for (a) the fcc lattice and (b) the bcc lattice (Smith, 1974).

It is important to distinguish between densely-packed planes in the CSL and densely-packed planes in the lattice. Frequently densely-packed planes in the CSL (which, by definition, have PCSD = 1) have relatively high indices and therefore do not coincide with densely packed planes in the lattice. The 111 plane in the Σ = 3 CSL is clearly an exception. Figure 3.6 illustrates this by showing two lattices in a $\Sigma = 3$ misorientation. We see that a boundary along 111 is densely-packed with respect to both the lattice and the CSL. The 112 plane is also a twin plane in the $\Sigma = 3$ system, and, although the PCSD = 1, the density of coincidence sites per unit GB area, σ, is lower, i.e. $0.82a^2$ compared with $2.31a^2$ for the 111 plane. There is experimental evidence for the dependence of GB properties on PCSD (Dimou and Aust, 1974).

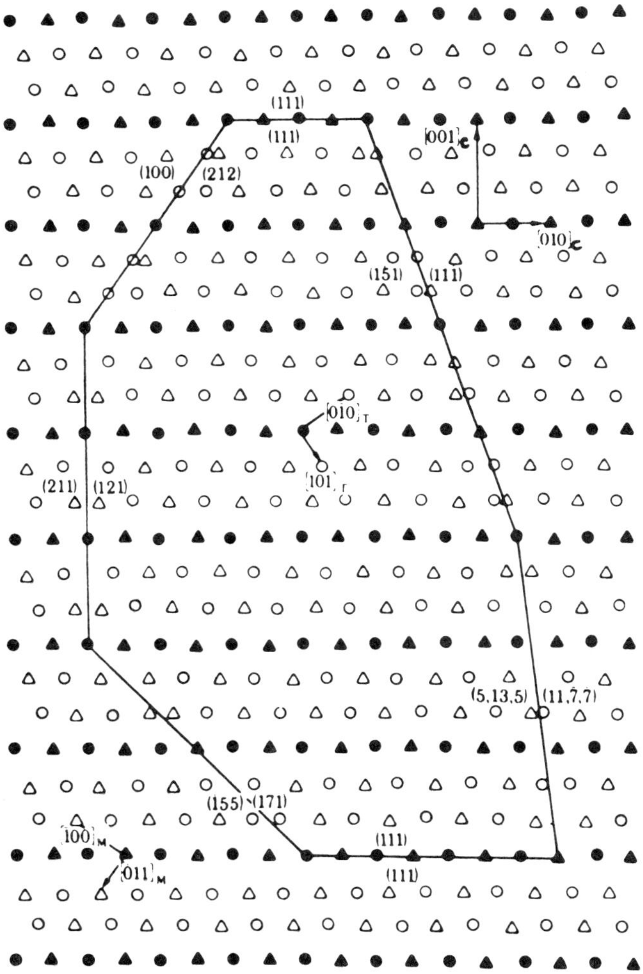

Figure 3.6 GBs in the $\Sigma = 3$ viewed along $01\bar{1}$. Triangles and circles represent different lattice layers and filled symbols represent coincidence sites. In two of the GBs every site is a coincidence site: these are the symmetrical twin GBs on 111 and 211 (Pond and Vitek, 1977).

3.2.5 Tilt and twist components in the coincidence site lattice

Where the GB plane normal in the crystal system of each grain (i.e. N_1 or N_2) is known in addition to Θ/UVW, the geometry of the GB can be decomposed to its tilt and twist components, as described in section 2.5. For CSLs this further characterisation of the GB geometry is pertinent because it provides information which can be related to the PCSD (Paidar, 1987). Using the interface-plane notation we will consider, with examples, four classifications: twist, symmetrical tilt, asymmetrical tilt GBs, and CSL GBs which are none of these. The interface-plane notation can, of course, also be used for non-CSL GBs.

A twist GB (TWGB) is characterised by the condition $N_1 = N_2$ and $\varphi \neq 0$. For example $[0\bar{1}2][012]$ 131.81°, as expressed in the interface-plane scheme, is a twist GB in the $\Sigma = 3$ system. All the GB planes for TWGBs in a particular CSL are given by *UVW* for each of the 24 symmetry-related solutions, and from equation 2.27a φ is given by the accompanying Θ. Thus another example of a TWGB in the $\Sigma = 3$ system is $[1\bar{1}0][110]$ 109.47°.

Tilt GBs may be categorised as symmetrical tilt (STGB) or asymmetrical tilt (ATGB). They can be recognised readily when the interface-plane scheme notation is used because $\varphi = 0$ (equation 2.27b). In addition, for STGBs the crystallographic form of N_1 and N_2 is equivalent. Several twist GBs can be generated on the same set of planes by varying φ; however for a particular set of planes there is only one STGB, and it is characterised by having the smallest planar unit cell for that plane set, hence the PCSD = 1 (Wolf, 1985). An example is $[111][11\bar{1}]$ 0° which refers to the 111 twin plane in the $\Sigma = 3$ CSL.

Where $\varphi = 0$ and N_1, N_2 are different but commensurate, the GB is an ATGB. A commensurate GB arises when the ratio of the planar GB unit cells (i.e. a projection of the 3-dimensional CSL unit cell onto the GB plane) is an integer. Since the area of the planar GB unit cell (denoted by Ar_1 and Ar_2 in grains 1 and 2 respectively) is proportional to the separation of planes parallel to the GB plane, we have:

$$Ar_1/Ar_2 = ((h_1^2+k_1^2+l_1^2)/(h_2^2+k_2^2+l_2^2))^{1/2} \qquad (3.10)$$

where $h_1k_1l_1$ and $h_2k_2l_2$ are the Miller indices of N_1 and N_2 respectively. For example {511} and {111} are commensurate planes because $Ar_1/Ar_2 = (27/3)^{1/2} = 3$. These planes constitute an ATGB in the $\Sigma = 3$ system which is illustrated on figure 3.6 along with some other examples of ATGBs and the two STGBs in the $\Sigma = 3$ system. Another example is [017][011] 0° in the $\Sigma = 5$ system.

For an example of a CSL which is neither a twist or a tilt GB, we will consider [110][543] 26.52°. For these planes, from equation 3.10 $Ar_1/Ar_2 = 5$ and the GB is a $\Sigma = 5$ CSL. However it is not a tilt GB because ψ is not zero and it is not a twist GB because N_1 and N_2 are not from the same 'family' of symmetry-equivalent planes. From the equations in section 2.5 we obtain Θ/UVW as 36.87°/100, which confirms that this GB is a $\Sigma = 5$ CSL. The twist axis, N_2, is [543] and the tilt axis and angle are respectively $[\bar{3}\bar{3}1]$ and 25.84°. The tilt and twist components for this GB were shown as spherical triangles in figure 2.10.

Asymmetric tilt GBs are ubiquitous in polycrystals, occurring far more frequently than STGBs or TWGBs (Merkle, 1988; Randle, 1989, 1991c; Merkle and Wolf, 1992). Table 3.5 is a list of combinations of N_1 and N_2 for N_1 = 100, 110, 111, 210 and 211 which constitute ATGBs for Σ up to 27. This Table, and another published elsewhere which lists ATGB combinations for N_1 = 100, 110 and 111 up to $\Sigma = 51$ (Sutton and Balluffi, 1987), can be used to compare with experimentally generated GB plane indices to check if they are near an ATGB condition. It should be noted that plane combinations are listed under the smallest Σ applicable, Σ_{min}, and can also occur in GBs having $\Sigma = n\Sigma_{min}$ where n is an integer. For example, a GB having planes 111/511 is listed in Table 3.5 under $\Sigma = 3$. However, this GB is also frequently observed in $\Sigma = 9$ GBs (Randle and Dingley, 1989, 1990a; Randle, 1991c).

There is some evidence that GB planes which are near densely packed lattice planes, i.e. 111, 110 or 100 in fcc materials, are important in polycrystals (Wolf, 1985, 1990b; Carter, 1988), even when only one of the two GB planes is near 111, 110 or 100. Certain GB plane combinations for ATGBs fulfil this criterion. For example, in the $\Sigma = 11$ system an ATGB having planes 111/19,1,1 is 4.3° from 111/100, and another with planes 311/39,35,35 which are 2.1° from 311/111. Other combinations are listed in Table 3.6 for Σ up to 27.

TABLE 3.5

ASYMMETRIC TILT GRAIN BOUNDARIES HAVING N_1 = {100}, N_2 = {hkl}

Σ	*hkl*	Σ	*hkl*	Σ	*hkl*	Σ	*hkl*
3	2 2 1	15	14 5 2	21	20 5 4	25	24 7 0
5	4 3 0	15	11 10 2	21	19 8 4	25	20 12 9
7	6 3 2	17	12 9 8	21	16 13 4	25	16 15 12
9	8 4 1	17	12 12 1	21	16 11 8	27	25 10 2
9	7 4 4	17	15 8 0	23	22 6 3	27	23 10 10
11	9 6 2	19	18 6 1	23	18 14 3	27	26 7 2
11	7 6 6	19	17 6 6	23	18 13 6	27	23 14 2
13	12 5 0	19	15 10 6			27	22 14 7
13	12 4 3						

ASYMMETRIC TILT GRAIN BOUNDARIES
HAVING N_1 = {110}, N_2 = {hkl}

Σ	*hkl*	Σ	*hkl*	Σ	*hkl*	Σ	*hkl*
3	4 1 1	15	16 13 5	21	25 16 1	25	31 17 0
5	7 1 0	15	19 8 5	21	23 17 8	25	25 24 7
5	5 4 3	15	20 7 1	21	20 19 11	25	29 20 3
7	9 4 1	17	23 7 0	21	29 5 4	25	27 20 11
7	8 5 3	17	17 15 8	23	28 15 7	25	35 4 3
9	8 7 7	17	21 11 4	23	24 19 11	25	31 15 8
9	11 5 4	17	20 13 3	23	25 17 12	25	32 15 1
11	12 7 7	17	24 1 1	23	21 19 16	25	28 21 5
11	13 8 3	19	17 17 12	23	32 5 3	27	23 23 20
11	15 4 1	19	23 12 7	23	31 9 4	27	35 13 8
13	17 7 0	19	24 11 5			27	28 25 7
13	13 12 5	19	25 9 4			27	29 19 16
13	16 9 1	19	21 16 5			27	37 8 5
13	15 8 7						

ASYMMETRIC TILT GRAIN BOUNDARIES
HAVING N_1 = {111}, N_2 = {hkl}

Σ	*hkl*	Σ	*hkl*	Σ	*hkl*	Σ	*hkl*
3	5 1 1	15	23 11 5	21	29 19 11	25	31 25 17
5	7 5 1	15	25 7 1	21	25 23 13	25	35 19 17
7	11 5 1	15	19 17 5	21	31 19 1	25	43 5 1
9	11 11 1	17	23 17 7	23	29 25 11	25	35 23 11
9	13 7 5	17	23 13 13	23	35 19 1	27	37 23 17
11	13 13 5	17	29 5 1	23	35 25 1	27	35 25 11
11	19 1 1	17	25 11 11	23	37 13 7	27	35 31 1
11	17 7 5	19	23 23 5				
13	17 13 7	19	29 11 11				
13	19 11 5	19	25 17 13				
		19	31 11 1				

ASYMMETRIC TILT GRAIN BOUNDARIES HAVING N_1 = {210}, N_2 = {hkl}

Σ	*hkl*	Σ	*hkl*	Σ	*hkl*	Σ	*hkl*
3	5 4 2	13	20 18 11	17	26 25 12	19	27 26 20
5	8 6 5	13	21 20 2	17	28 25 6	19	30 28 11
5	10 4 3	13	22 19 0	17	30 17 16	19	30 29 8
5	11 2 0	13	24 13 10	17	30 23 4	19	35 18 16
7	10 9 8	13	26 12 5	17	31 22 0	19	35 24 2
7	12 10 1	13	27 10 4	17	32 15 14	19	36 22 5
7	15 4 2	13	28 6 5	17	33 16 10	19	37 20 6
9	16 10 7	13	29 2 0	17	34 15 8	21	34 32 5
9	17 10 4	15	23 20 14	17	36 10 7	21	38 20 19
9	20 2 1	15	25 22 4	17	38 1 0	23	31 30 28
11	18 16 5	15	26 20 7			23	34 33 20
11	19 12 10	15	31 10 8			23	38 25 24
11	20 13 6	15	32 10 1			23	39 32 10
11	20 14 3						
11	21 10 8						
11	24 5 2						

ASYMMETRIC TILT GRAIN BOUNDARIES HAVING N_1 = {211}, N_2 = {hkl}

Σ	*hkl*	Σ	*hkl*	Σ	*hkl*	Σ	*hkl*
3	5 2 2	11	19 14 13	15	25 23 14	19	29 29 22
3	7 2 1	11	19 19 2	15	25 25 7	19	31 26 23
5	10 7 1	11	23 14 1	15	26 25 7	19	34 29 13
5	11 5 2	11	25 10 1	15	29 22 5	19	34 31 7
7	13 10 5	11	26 5 5	15	31 17 10	19	35 29 10
7	13 11 2	11	26 7 1	15	34 13 5	19	37 26 11
7	17 2 1	13	22 19 13	15	35 11 2	21	34 31 23
9	14 13 11	13	23 17 14	17	25 25 22	21	35 35 14
9	17 14 1	13	23 22 1	17	26 23 23	21	37 34 11
9	19 10 5	13	25 17 10	17	31 22 17	21	38 29 19
9	19 11 2	13	26 17 7	17	34 23 7	23	38 37 19
9	22 1 1	13	29 13 2	17	35 22 5	25	37 35 34
		13	31 7 2	17	37 14 13		
				17	37 19 2		
				17	38 13 11		
				17	38 17 1		

TABLE 3.6

ASYMMETRICAL TILT GBs HAVING ONE OF BOTH PLANES WITHIN A SMALL ANGLE V° OF 111, 100 or 110.

(* = 111, + = 100, ^ = 110)

Σ	planes	ν°	Σ	planes	ν°
7	210/10 9 8	1.5*	17	211/25 25 22	3.4*
			17	211/26 23 23	3.4*
9	110/877	3.7*	17	210/38 1 0	1.5+
9	111/11 11 1	3.7+	17	110/24 1 1	3.4+
			17	100/12 12 1	3.4^
11	100/766	4.3*	17	221/36 36 3	3.4^
11	221/20 20 17	4.3*	17	221/38 34 1	3.4^
11	311/23 21 19	4.5*	17	310/39 37 0	1.5^
11	111/19 1 1	4.3+			
11	211/19 19 2	4.3^	19	310/36 35 33	2.1*
11	310/25 24 3	5.1^	19	310/39 35 35	3.0*
13	210/21 20 2	4.2^	21	221/38 37 34	2.7*
13	210/22 19 0	4.2^			
13	210/29 2 0	3.9^	23	210/31 30 28	2.4*
13	211/23 22 1	2.2^			
13	221/29 26 2	4.3^	25	211/37 35 34	2.0*
13	310/31 27 0	3.9^			
			27	110/23 23 20	3.7*
15	310/35 32 1	2.8^	27	111/35 31 1	3.7^

3.3 OTHER MODELS OF GRAIN BOUNDARY GEOMETRY

In this section some other models pertaining to GB geometry are briefly described.

3.3.1. The coincident axial direction/planar matching model

The coincident axial direction (CAD) and planar matching (PM) models are essentially equivalent. A CAD/PM GB (which will be more concisely referred to as a CAD) is characterised by the near parallelism of the same family of planes from each abutting grain, or the equivalent approach is to consider the near coincidence of the normals to these planes (Warrington and Boon, 1975; Ralph, 1975; Watanabe, 1983). Originally the PM theory evolved to account for the observation of intrinsic dislocations in GBs which were outside the CSL range (Pumphrey, 1976).

Compared to the CSL model, the CAD model is associated with one-dimensional lattice matching whereas the CSL model is essentially three-dimensional. Consequently, the CAD model is a relatively poor predictor of special GB geometry (Dechamps *et al,* 1988). However, it is still sometimes applicable to situations where a particular misorientation axis is especially prevalent in a polycrystal, e.g. where there is a strong fibre texture (Randle and Ralph, 1988c). As the Σ-value increases, the shape of the CSL unit cell becomes increasingly anisotropic and in the limit will approach the interplanar spacing in one direction and zero in the other two (Grimmer *et al,* 1974). In other words it approaches the unidimensional form associated with a CAD. Thus it is justifiable to consider high Σ CSLs in terms of the CAD model. Furthermore, it may be applicable to analyse the crystallographic geometry of multiple grain junctions in terms of the CAD model, since the junction itself is a one-dimensional defect (Palumbo and Aust, 1990b).

The angular range over which CADs exist is predicted in a manner analogous to that for a CSL. Similarly to equation 3.8 for the CSL case, the expression for the angular limit of a CAD is

$$v_c = v_0(a/\boldsymbol{b})\pi^{\frac{1}{2}} \tag{3.11}$$

where v_0 has the same meaning as in equation 3.8, a is the lattice parameter, $\boldsymbol{b}$ is the Burgers vector of primary dislocations in a low angle GB and π is $h^2+k^2+l^2$ where *hkl* are the indices of the matching planes/coincident direction which corresponds to planes with a finite structure factor for fcc and bcc materials. For fcc and bcc materials the CAD limit is 1.35 and 1.10 times greater respectively than the CSL limit. Table 3.7 gives values of v_c for $\pi \leq$ 24 in fcc materials.

TABLE 3.7

MAXIMUM ANGULAR DEVIATIONS v_c ° FOR CAD MISORIENTATIONS (FCC)

Σ	*hkl*	v_c °
3	111	11.7
4	200	10.1
8	220	7.2
11	311	6.1
19	331	4.7
20	420	4.5
24	422	4.1

It is predicted that 56% of GBs in a randomly oriented fcc polycrystal should be CADs having $\pi < 8$, i.e. 111, 200 and 220 matching planes compared to 9% for CSLs (Warrington and Boon, 1975) (see also section 6.2).

3.3.2 The O-lattice

The O-lattice is defined as the point set of all possible origins for rotational transformations between the lattices in a dichromatic pattern (Bollmann, 1970). In other words it is a set of *all* point coincidences, whether or not they are lattice points. Hence the CSL, which consists of lattice points, is a subset of the O-lattice. The basic equation which defines an O-lattice vector is

$$(\mathbf{I} - \mathbf{A}^{-1})\, x^o = b^L \tag{3.12}$$

where **I** is the identity matrix, **A** the transformation matrix between the two lattices, $\boldsymbol{x^o}$ an O-lattice vector and $\boldsymbol{b^L}$ a corresponding lattice vector. Details of how the O-lattice equation is used can be found elsewhere (Bollmann, 1970; Ralph *et al*, 1981; Gleiter, 1982). Associated with the O-lattice is the displacement shift complete (DSC) lattice which is used for describing intrinsic GB dislocations (Grimmer *et al*, 1974). The O-lattice is the most complete and elegant model for describing GB geometry. However, for GB characterisation of large datasets in polycrystals, the CSL model is appropriate and satisfactory, and hence widely used (see chapter 7).

3.3.3 Structural unit models

An alternative view of GB structure to that discussed so far is to focus on the spaces or polyhedral units made by the surrounding configurations of atoms in a GB. These structural units in turn give rise to a repeating arrangement of them in the GB plane, where this is rational (Ashby *et al*, 1978; Pond *et al*, 1979). Where only one type of unit is involved, the GB is referred to as favoured; where the structural unit period is composed of more than one unit type the GB is unfavoured (Sutton and Vitek, 1983). Any CSL boundary can be described as combinations of the basic units, and the structural units can equivalently be described in terms of dislocations. The structural unit approach is not amenable to experimental investigation except by high resolution microscopy of bicrystals. It is included here only to show the link with the GB geometry.

3.4 MULTIPLE JUNCTIONS IN POLYCRYSTALS

In polycrystals two grains meet at a GB, which is a surface. Three or more grains can only meet at a line, rather than a surface. Usually the number of grains which meet in a line is three and the line is then known as a triple line or a triple grain junction. Energy balance considerations require that, for GBs which are in equilibrium and have equal energies the dihedral angle between each grain at the triple junction is 120°. GBs which have particularly low energies alter this balance, (Watanabe, 1987; Randle 1990b) which also make it possible to accommodate more than three GBs at a grain junction, which is then called a multiple grain junction (Kopezky *et al,* 1991).

A second consideration for multiple junctions, which is of particular interest in the context of the crystallography of GBs, is the combination of misorientations which occurs (Don and Majumdar, 1986; Doni and Bleris, 1983; Randle, 1990c, 1991b). Three misorientation matrices, $\mathbf{M}_1,\mathbf{M}_2,\mathbf{M}_3$, which relate to three GBs respectively at a triple junction, must obey the following condition:

$$\mathbf{M}_1\mathbf{M}_2\mathbf{M}_3 = \mathbf{I} \qquad (3.13a)$$

or alternatively

$$\mathbf{M}_3 = \mathbf{M}_1\mathbf{M}_2^{-1} \qquad (3.13b)$$

where the three matrices are expressed in the same reference system. A diagram of a triple grain junction is shown in figure 3.7 which refers to a Σ = 3, Σ = 27a and Σ = 9 CSL between grains AB, BC and CA respectively. A further example which illustrates a possible relationship at a triple grain junction is

$$131.81°/210 \quad + \quad 48.19°/210 \quad = \quad 180°/210 \qquad (3.14a)$$

$$\Sigma = 3 \quad + \quad \Sigma = 15 \quad = \quad \Sigma = 5 \qquad (3.14b)$$

In equation 3.14a we have substitued the appropriate Θ/UVW for the misorientation matrix, because this makes clear a further three rules which govern the crystallographic characteristics of triple junctions:

1. They share a common misorientation axis which is 210 in this case;
2. The sum of two of the angles of misorientation is the third, i.e. 48.19 + 131.81 = 180;
3. The product or quotient of two of the Σ-values gives the third, i.e. 3 and 5 → 15.

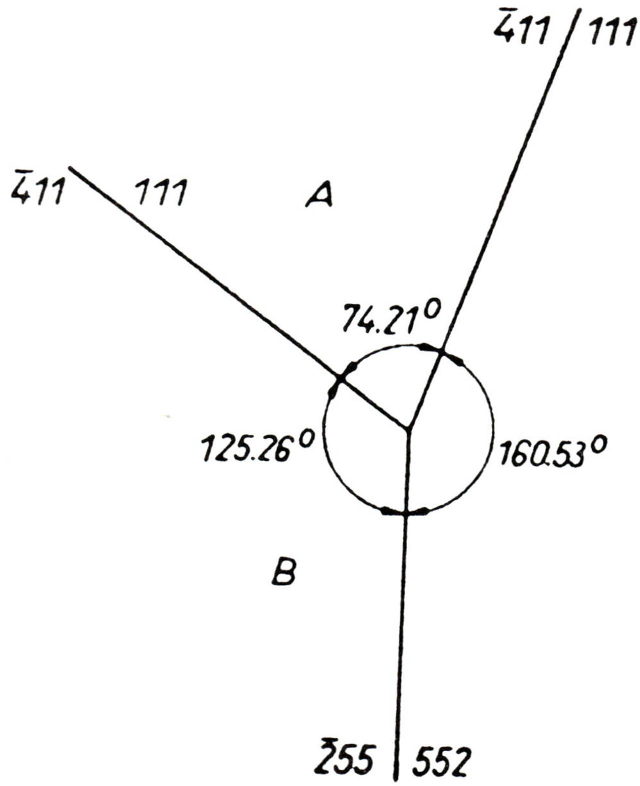

Figure 3.7 Schematic illustration of a triple GB junction comprising a Σ = 3 (AB), Σ = 27a (BC) and Σ = 9 (CA) (Doni and Bleris, 1988).

Thus equation 3.14 leads to the conditions

$$(UVW)_3 = (UVW)_2 = (UVW)_1 \tag{3.15a}$$

$$\Sigma_3 = \Sigma_2\Sigma_1 \tag{3.15b}$$

$$\Theta_3 = \Theta_2\Theta_1 \tag{3.12c}$$

When obtaining the third misorientation from a knowledge of two others at a triple junction, it is necessary to take into account the clockwise or anticlockwise sense of the combination, i.e. the angles (or Σ-values) can be added or subtracted. Therefore a combination of a $\Sigma = 3$ and a $\Sigma = 15$ CSL will result in a third GB which is either a $\Sigma = 5$ CSL as described above, or a $\Sigma = 45$ CSL:

$$131.81°/210 \quad - \quad 48.19°/210 \quad = \quad 83.62°/210 \qquad (3.16a)$$

$$\Sigma = 3 \quad - \quad \Sigma = 15 \quad = \quad \Sigma = 45b \qquad (3.13b)$$

Relationships of the type shown in equations 3.14 to 3.16 are clearly of importance in polycrystals. The most common occurrence is of triple junctions having $\Sigma = 3^n$, e.g. a junction of two $\Sigma = 3$ and one $\Sigma = 9$ GB, formed either when a $\Sigma = 9$ CSL dissociates into two $\Sigma = 3$ GBs or when two $\Sigma = 3$s combine to form a $\Sigma = 9$ as illustrated in figure 3.7 (Goodhew *et al,* 1978; Garg *et al,* 1989). An experimental example which illustrates these types of triple line relationships is shown in figure 6.15.

Triple junctions can also be interpreted as line defects whose structural characteristics depend upon the balance of dislocations from each GB in the triple junction. It is possible to calculate this balance by a procedure whose details are beyond the scope of this book but, are described elsewhere (Bollmann, 1984, 1989). Briefly, the procedure involves redefining the three misorientations at the junction in terms of the coordinates of one of the adjoining lattices. The misorientations can then be converted to 'nearest neighbour relationships', i.e. the most widely spaced GB dislocation array in the GB, via a transformation **U**. At a triple junction comprising GBs a,b,c (figure 3.8a) we can then characterise the junction by a tensor **T**:

$$\mathbf{T} = \mathbf{U}_a\mathbf{U}_b\mathbf{U}_c \qquad (3.17)$$

When **T** is the identity, the dislocation balance is satisfied at the junction and this is called an **I**-line. Conversely, a **U**-line results when **T** is not the identity and thus the dislocation content at the junction is not balanced (figure 3.8b).

I-lines and **U**-lines are important because they have been found to affect the properties of GB junctions (Palumbo *et al,* 1990; Palumbo *et al,* 1991). For example, it has been shown that **U**-lines are preferential sites for segregation or corrosion (Palumbo and Aust, 1989) and it is expected that GB triple line characterisation might also play be important in such phenomena as grain growth and crack nucleation (Bollmann, 1989).

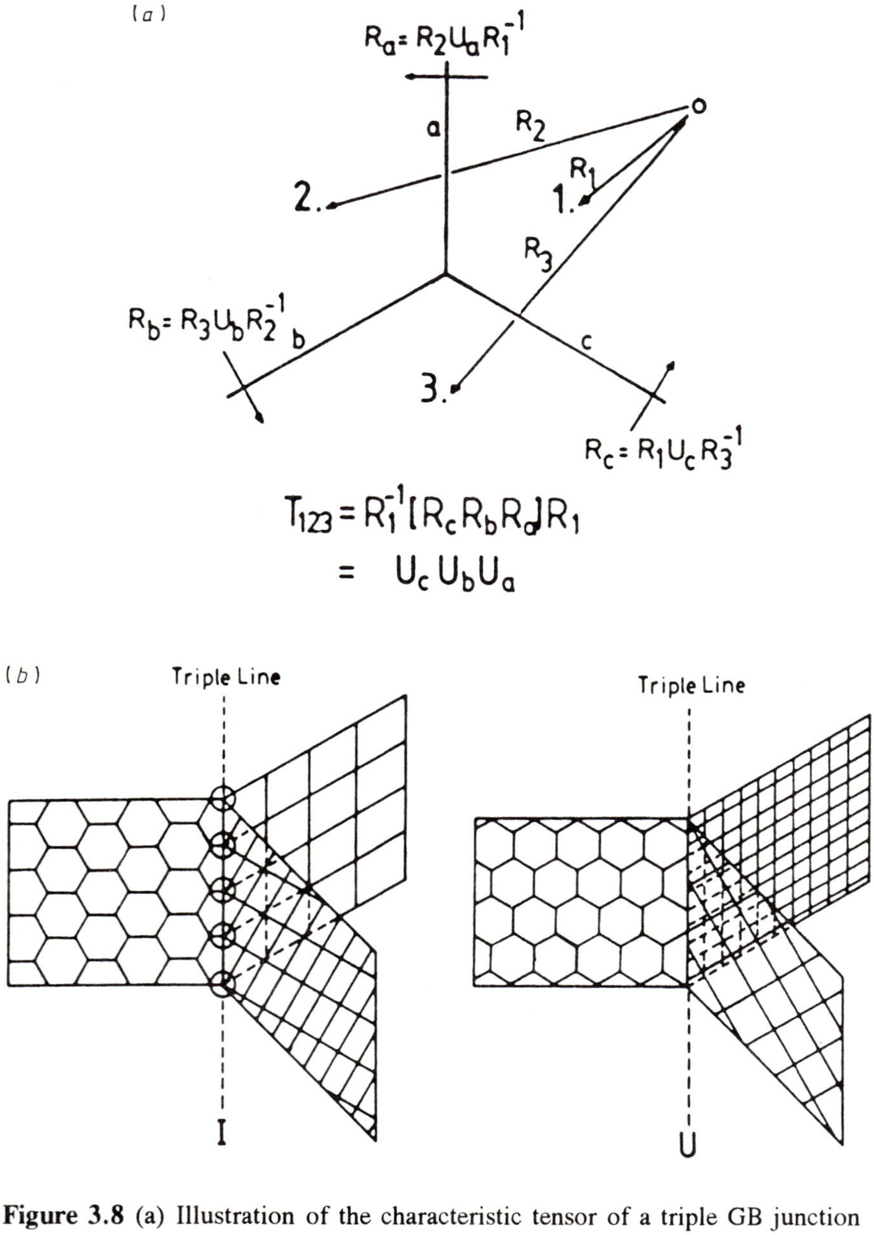

Figure 3.8 (a) Illustration of the characteristic tensor of a triple GB junction from the **U** matrices (nearest neighbour relationships), calculated as shown from the misorientations . (b) Illustration of the distinction between an **I**-line and a **U**-line. For the **I**-line the adjoining GB dislocation arrays are balanced (Palumbo and Aust 1990b).

4

EXPERIMENTAL METHODS

4.1 INTRODUCTION

In this chapter we explain how to obtain 'raw' input data for GB analyses. Almost always the crystallographic part of these input data are Kikuchi electron diffraction patterns. A technique which is based on electron microscopy is the optimal method for the study of GB geometry because the probe can be made small enough to sample a volume which has a homogeneous or near homogeneous orientation: in other words the sample volume is less than an individual grain volume or, in some cases, the subgrain volume (Randle, 1992c). This is in contrast to X-ray diffraction where many grains are included in the sample volume. An exception to sampling of a single, discrete orientation by an electron beam technique is if the crystal is imperfect over a region of the order of a few nanometers, i.e. smaller than the beam size. This situation is encountered in highly cold worked materials and causes the Kikuchi lines in the diffraction pattern to broaden and become diffuse.

A further advantage of electron microscopy for GB geometrical work is that an *image* of the precise location of the diffracting volume together with the surrounding microstructural environment is available. Thus the crystallographic measurements can be combined with spatial/stereological data (Schwarzer and Weiland, 1988). The (mis)orientational/spatial correlation is referred to again in section 6.5

Table 4.1 contains a list of techniques for measuring GB geometries. Details of these techniques can be found elsewhere as follows: X-ray techniques and macrotexture (Wenk, 1985; Bunge, 1987); synchrotron radiation (Gastaldi *et al,* 1988); electron-beam based methods (Humphreys, 1988; Schwarzer, 1990); SEM-based diffraction (Dingley, 1981); TEM-based orientation measurement (Schwarzer and Weiland, 1988); EBSD and neutron diffraction (Juul Jensen and Randle, 1989); microtexture determination (Randle, 1992c) and EBSD (Dingley and Randle, 1992). Methods which use X-rays to probe the microstructure are sometimes still used for GB analysis, but as indicated in Table 4.1, the disadvantages of X-ray based methods are that only very large grains can be sampled individually (unless a synchrotron source is used) and also there is a lack of imaging facilities. Also, the availability of specific equipment is an important consideration. Whereas an X-ray diffractometer used to be a more common laboratory instrument than an electron microscope, today most research laboratories are either equipped with SEM and/or TEM or have access to them.

TABLE 4.1

SUMMARY OF TECHNIQUES FOR MEASURING GRAIN BOUNDARY GEOMETRY

Technique	Status	Resolution
X-ray (Laue)	Largely superceded by electron beam techniques.	1 mm
X-ray (synchrotron)	Not convenient for large-scale investigations - low availability.	10 μm
SEM (SAC)	Largely superceded by EBSD.	10 μm
SEM (EBSD) investigations.	Optimum technique for most	0.5 μm
TEM (SAD)	Largely superceded by CBED.	100 nm
TEM (CBED)	Optimum technique for deformed materials.	10 nm
TEM (HVEM)	Optimum TEM technique - greater volumes are electron-transparent.	10 nm

As far as TEM is concerned, the input data for measuring GB parameters can be obtained using any modern instrument. No additional facilities are needed because formation of a diffraction pattern is an inherent part of the image formation process in the TEM. On the other hand, a diffraction capability is not a feature of SEM because the image is formed in a different way to TEM (Hirsch *et al,* 1965). Consequently non-standard equipment must be added to the SEM before crystallographic work can be performed. The two SEM techniques which provide diffraction information are selected area channelling (SAC) and electron back-scatter diffraction (EBSD). SAC has now largely been superceded by EBSD, and so we will concentrate only on the latter.

The primary input for GB geometry analysis is a Kikuchi electron diffraction pattern from each of two adjoining grains. Kikuchi patterns are described and discussed in section 4.2. The orientation of each grain is obtained from them and then the misorientation is computed as described in section 5.2. The inclusion of other GB parameters within the scope of a particular investigation, such as the GB plane normal or the specification of external reference axes, require additional, spatial measurements to be made. It is important to realise that any electron diffraction technique, whether TEM or SEM based, produces Kikuchi diffraction patterns which are essentially equivalent.

The general operation of a TEM/SEM and standard specimen preparation, both of thin foils for TEM and metallographic polishing and etching for SEM, will not be described in this book and details can be found elsewhere (Chescoe and Goodhew, 1990). There are a few specimen preparation requirements which relate particularly to EBSD and these are mentioned in section 4.4.1. The layout of the remainder of this chapter, which deals with the experimental side of GB geometry analysis, is as follows. In section 4.2 the formation and characteristics of Kikuchi diffraction patterns are described in detail, since these are the basic input for any subsequent crystallographic analysis. Experimental details concerning the collection of GB data in the TEM and SEM are described in sections 4.3 and 4.4 respectively. In the present chapter we deal only with the actual collection of raw data; the subsequent processing of this data is described in chapter 5.

4.2 KIKUCHI ELECTRON DIFFRACTION PATTERNS

In very thin TEM specimens a single crystal diffraction pattern takes the form of the classical 'spot' pattern which is associated with selected area diffraction (SAD). These patterns are not suitable for the accurate measurement of orientation because the positions of the reflections (spots) may not change until the specimen has been tilted through several degrees (Hirsch *et al,* 1965). If, however, the diffraction pattern arises from thicker crystal such as a relatively thick TEM foil or an SEM specimen, or alternatively if a microdiffraction TEM technique is used, e.g. convergent beam electron diffraction, CBED (Ralph and Ecob, 1984), an additional diffraction effect occurs which can be exploited to measure the orientation of the crystal very accurately. The effect is Kikuchi electron diffraction and is a consequence of the scattering of the incident electron beam in all directions during its penetration of the thick specimen. This means that there is always a proportion of scattered electrons which impinge on all lattice planes at the Bragg angle; these electrons are then diffracted according to Bragg's law:

$$2d \sin \Theta_B = n\lambda \tag{4.1}$$

where d is the interplanar spacing for a family of planes, λ the electron wavelength, n the order of reflection and Θ_B the Bragg angle. Bragg diffraction occurs from the planes which are on both sides of the scattering source, giving rise to two cones of electron radiation for each family of lattice planes.

Figures 4.1a and b show Kikuchi diffraction in two and three dimensions respectively for a single set of lattice planes. If a flat surface such as a screen or piece of photographic film is placed so as to intercept the cones of diffracted radiation, each family of planes in the crystal is represented by a pair of conic sections. The conic sections approximate to pairs of straight parallel lines rather than curves because in reality Bragg angles are much smaller than shown on the schematic representation in figure 4.1, typically about 0.5°, and so the apex angle of the cones is extremely large. The line pairs, which are known as Kikuchi lines, consist of a line which is darker than the background and one which is brighter. This is due to the unequal transfer of electrons along each cone surface. The darker cone arises because fewer electrons are scattered in this direction since it makes a larger angle of scattering with the primary electron beam, and vice versa for the brighter line. This is illustrated on figure 4.1b. Only electrons which are elastically scattered through the Bragg angle contribute to the Kikuchi lines; inelastically scattered electrons form a diffuse background to the pattern.

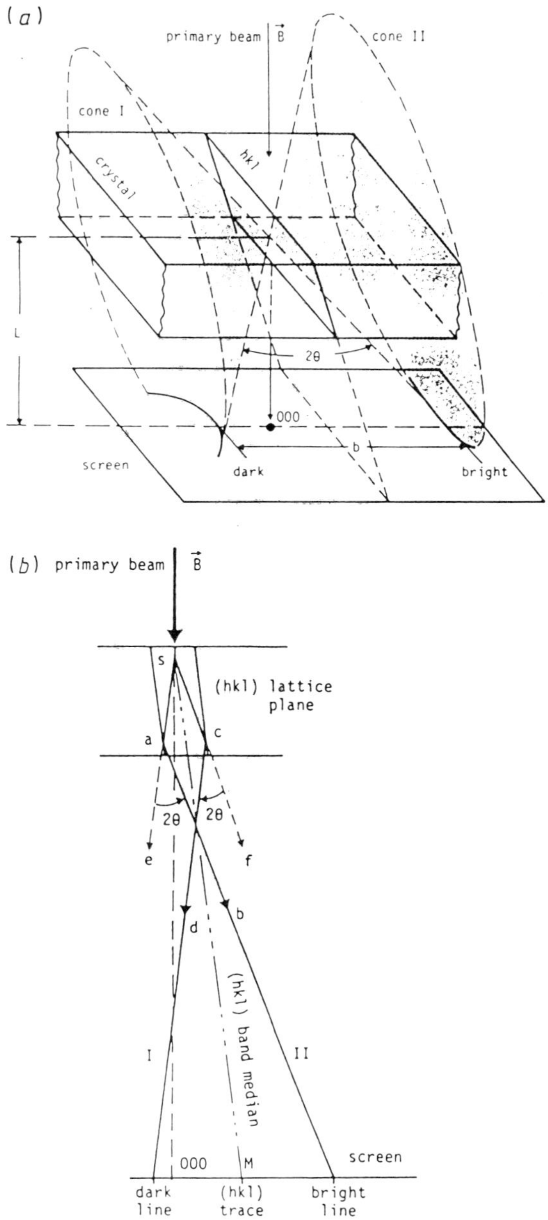

Figure 4.1 Schematic illustration of the formation of Kikuchi line pairs during diffraction. (a) Three-dimensional view showing that Kikuchi lines arise from cones of diffracted electrons. (b) Two-dimensional view showing the formation of a dark and bright line (Schwarzer and Weiland, 1988).

Essentially, a Kikuchi pattern is a projection of the geometry of all the lattice planes in a crystal. In other words it provides a very precise 'map' of the distribution of the lattice plane geometry. From Bragg's law the Kikuchi line spacing represents $2\Theta_B$, and the actual trace of the plane occurs midway between a Kikuchi line pair. Important directions in the crystal, zone axes, occur where several planes intersect and appear on a Kikuchi patterns as intersections of Kikuchi line pairs. Figure 4.2a and b are typical patterns from an fcc and bcc material respectively. These patterns include two stereographic unit triangles in each, which are marked for clarity. The zone axes at the corners of these unit triangles, 001, 011, 111 and 101, and also some other prominent zone axes, are labelled. Hence any experimentally obtained Kikuchi pattern from an fcc or bcc phase can be identified by matching it to part of the patterns in figure 4.2.

When interpreting the Kikuchi pattern, it is important to remember that the rules which govern symmetry and absences of planes with certain indices are obeyed. Hence, for example, the 001 axis in the Kikuchi patterns in figure 4.2 can be recognised by its four-fold symmetry and also we see that, because of systematic absences of certain reflections, the Kikuchi line spacings are different for patterns arising from fcc and bcc crystals. For example, a detailed study of these features can be used to identify many point groups. However, this application is beyond the scope of this book because here only cubic materials are considered. Further information concerning the information contained within Kikuchi patterns (particularly EBSD patterns) can be found elsewhere (Baba-Kishi, 1986; Baba-Kishi and Dingley, 1989). Usually, the operator knows the crystal structure of the specimen before the crystallographic data is collected. If this is not the case a correct identification of the Bravais lattice, that is, if it is fcc or bcc, can be made from the relevant structure factors or more simply and directly by visually matching the experimental pattern to figure 4.2a or b.

In practical terms a Kikuchi pattern is a map of angular relationships between directions and planes in the crystal. Linear distances measured on the pattern represent angles; for example the distance measured between the projected poles 001 and 011 represents the interzonal angle 45°. Interplanar angles can also be obtained from the Kikuchi pattern but this is not of relevence to orientation measurement. The projection geometry of the Kikuchi pattern is related to the real geometry of planes in the crystal via the gnomonic projection (McKie and McKie, 1974). This projection is based on the reference sphere concept; the projection point is the centre of the reference sphere and the projection plane is tangential to the north pole as illustrated in figure 4.3. The gnomonic projection can be compared to the more familiar stereographic projection where the projection point and plane are the south pole and equatorial plane respectively.

In the formation of a Kikuchi pattern a pole P (zone axis) which is at an angle τ from the north pole N projects gnomonically at a distance $r\tan\tau$ from

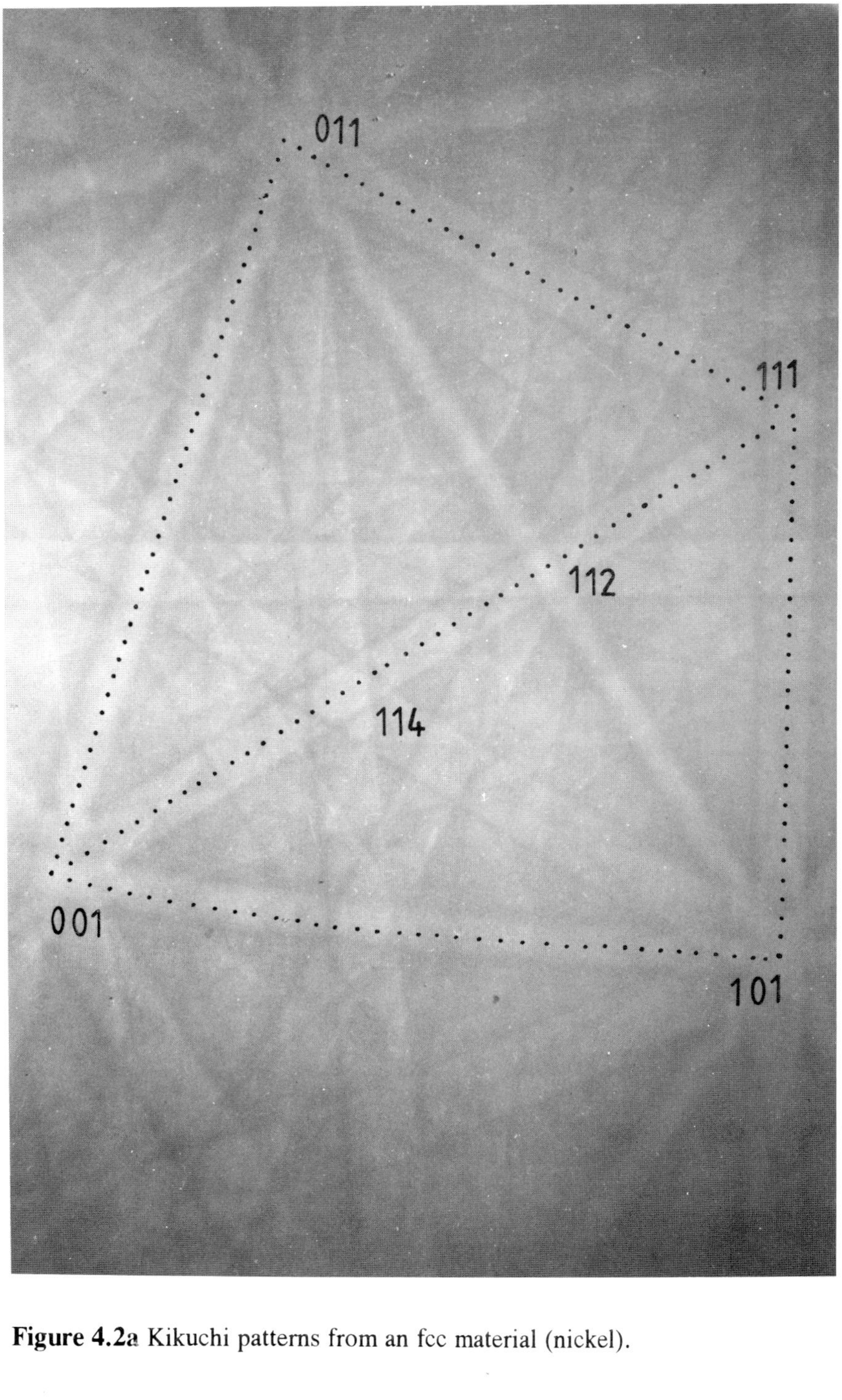

Figure 4.2a Kikuchi patterns from an fcc material (nickel).

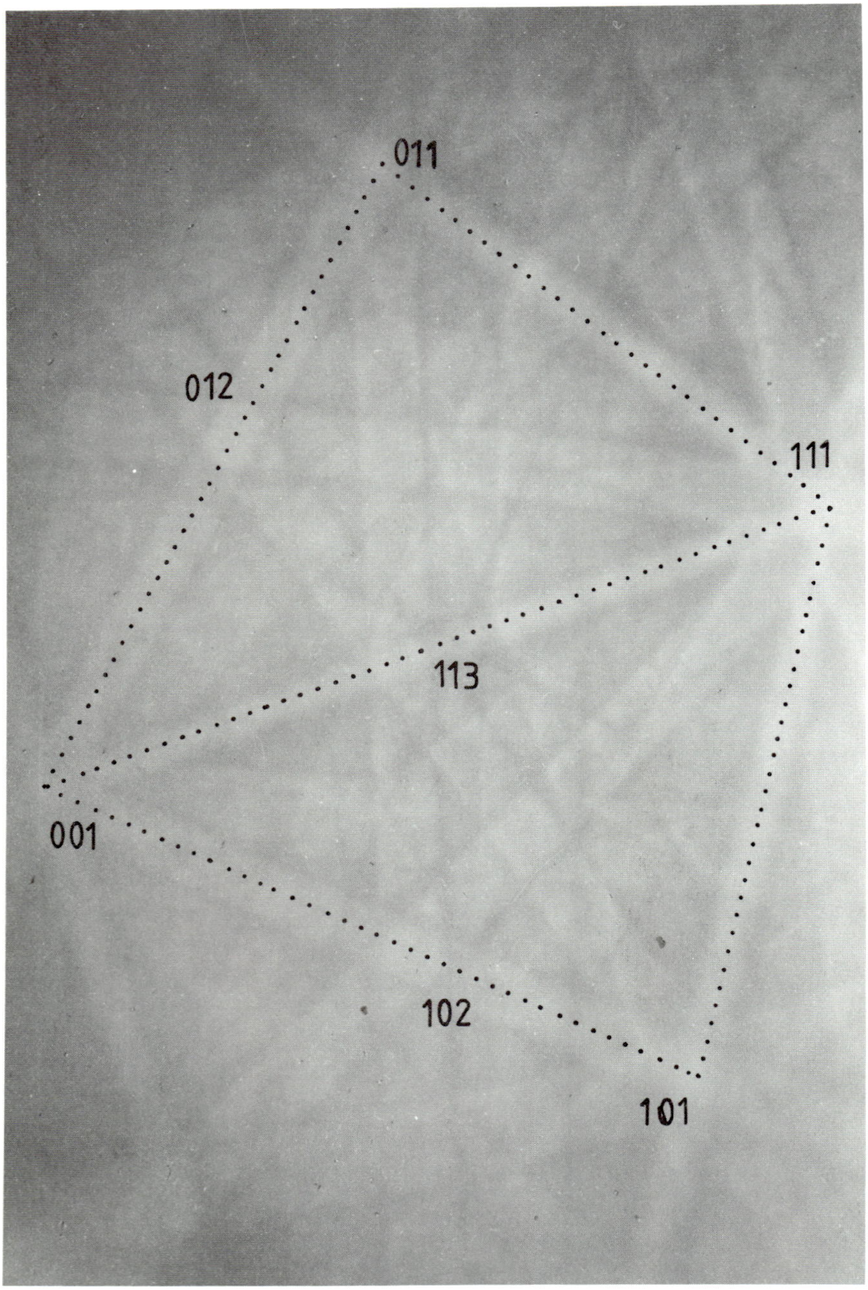

Figure 4.2b Kikuchi patterns from a bcc material (iron). The patterns were obtained by EBSD (Dingley *et al*, 1992).

N, where *r* is the radius of the sphere as shown on figure 4.3. The point *N* represents the centre or origin of the Kikuchi pattern and *r* represents the 'camera length', i.e. the specimen to screen distance. These two parameters are important when extracting orientation measurements from Kikuchi patterns and will be referred to again in sections 4.4.1 and 5.2. Since the relationship between τ and the linear distance on the pattern is a function of tanτ, distortion is introduced into a Kikuchi pattern which becomes visually obvious for large values of τ. This can be apparent in EBSD patterns because an angular range of up to 80° is covered (see section 4.4).

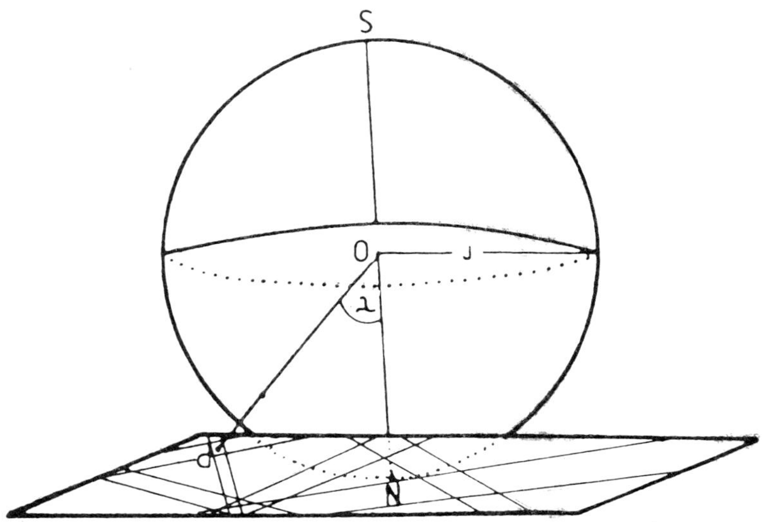

Figure 4.3 Schematic illustration of the gnomonic projection. The point *N* represents the pattern centre and *P* is a pole in the diffraction pattern. The distance between *P* and the pattern centre is given by rtan τ.

4.3 DATA COLLECTION IN THE TEM

4.3.1 Grain misorientations in the TEM

The optimal method used in the TEM to formulate a Kikuchi pattern is microdiffraction. The procedure to obtain patterns is to focus a small probe onto the selected region in the image and switch to diffraction mode. The pattern is then either photographed for off-line analysis or alternatively it may be analysed on-line in which case recording the pattern is not necessary. Kikuchi lines also appear in SAD patterns from thick crystal, and although the lines are not so well-defined as in a microdiffraction pattern they can still be used for orientation determination. Whether microdiffraction or SAD is used, two Kikuchi patterns (one from each neighbouring grain) are sufficient information to compute a misorientation. An image of the region from which diffraction information has been obtained is usually recorded, or several images if appropriate. This allows the incorporation of stereological data e.g. projected GB area and/or grain size (see section 6.5).

Figure 4.4 shows an example of a typical microdiffraction pattern pair which would be used to obtain an orientation. The patterns were obtained at an accelerating voltage of 120kV and the camera length was 210mm. Many modern TEMs can now be operated at an accelerating voltage of up to 300kV. It is advantageous to use as high an accelerating voltage as possible because this allows greater thicknesses of foil to be penetrated and thus extends the range of the specimen which can be used for analysis. High voltage electron microscopes (HVEM) can be operated at up to 1MV. At this accelerating voltage aluminium specimens about 10μm thick are electron transparent and for a material with atomic number near that of iron or copper the value is about 2μm (Berger *et al*, 1988). However, there are relatively few HVEMs available and most GB characterisations are carried out using an accelerating voltage in the range 100-300 kV.

With regard to the selection of a suitable camera length, the camera length which was used to generate the diffraction patterns shown in figure 4.4 was 210mm. This allows the maximum angular range of the pattern to be covered for an accelerating voltage of 120kV, that is, about 21°. However, a longer camera length gives a more magnified pattern but less angular range. A suitable camera length is the longest one where there are still sufficient zone axes in the pattern to facilitate its unambiguous recognition and indexing. Three zone axes is the minimum requirement.

Once the Kikuchi pattern has been obtained on the microscope screen it is either photographed for off-line analysis or interrogated *in situ*. The objective for on-line analysis is to transmit the coordinates of either three operator-selected zone axes or two points on at least three Kikuchi line pairs to the computer. The coordinates of the central beam and a reference frame are also needed. The required coordinates are obtained by driving the diffraction spot alignment potentiometers until a selected position coincides with a reference mark, e.g. the beam stop. The corresponding voltage drop over the post-specimen beam deflection coils of the microscope are passed to the computer using an AD converter, which indicates the selected position (Schwarzer, 1990). The principles of extracting an orientation from a Kikuchi pattern are described in section 5.3 with relevant examples.

Sometimes the GB parameters need to be known relative to external reference axes in the specimen. For such cases the external reference axes have to be marked on each TEM specimen disc prior to preparation, and then the reference directions have to be aligned carefully in the microscope itself. This procedure inevitably introduces some error in relating the misorientation and/or plane to the external axes.

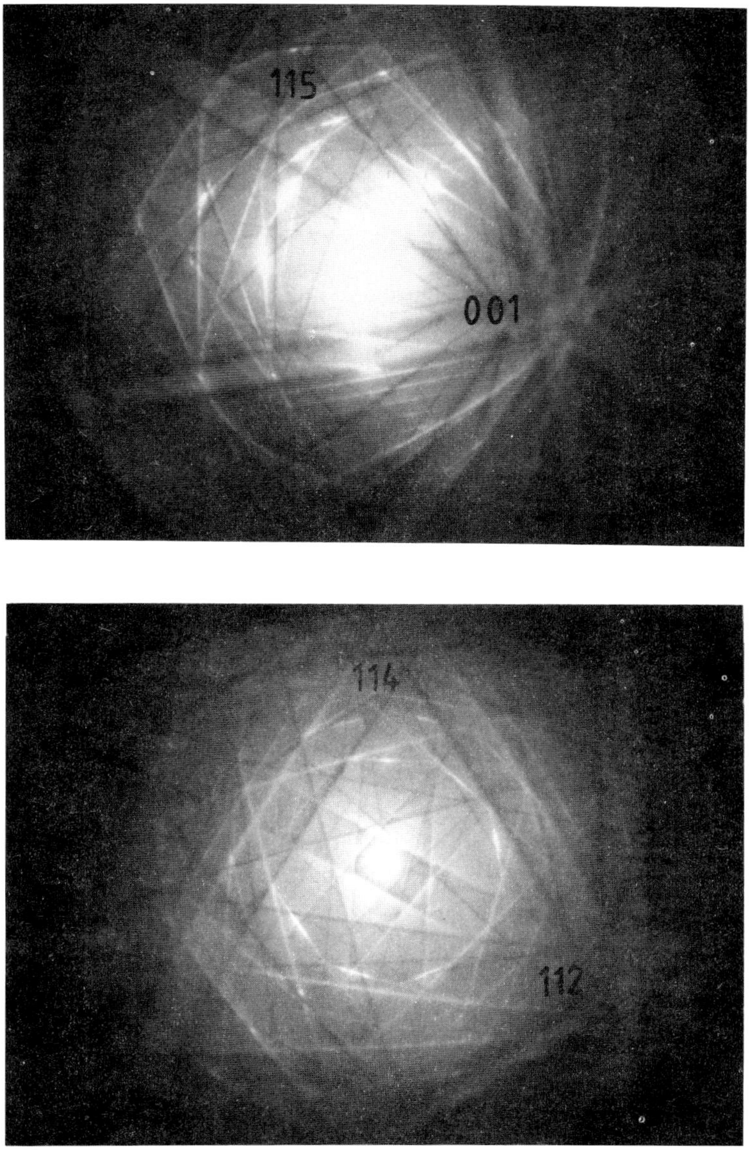

Figure 4.4 Diffraction patterns, obtained using CBED in a TEM, from either side of a GB in a nickel-based (fcc) alloy (Randle and Ralph, 1986).

4.3.2. Grain boundary planes in the TEM

In addition to the three degrees of freedom associated with the misorientation, the other two intrinsic degrees of freedom may be included as parameters in the GB analysis, i.e. the direction of the GB plane normal. To measure the direction of the GB plane normal, N, in a TEM specimen, we need a Kikuchi pattern from each neighbouring grain as described in the previous subsection, and also the inclination of the GB within the foil. The angle of inclination, φ, is usually expressed as the angle between the beam direction and the GB plane as illustrated in figure 4.5a. Once this angle is known, N can be determined relative to the orientation of each neighbouring grain and also to the external axes if they are known.

There are several different methods documented to obtain the angle of inclination and thus determine N. The most simple experimental method is to make φ zero, i.e. to tilt the foil until the GB plane is parallel to the beam direction. In other words the projected GB width becomes a minimum in the image, and the only information required to obtain N is a diffraction pattern from each interfacing grain and the direction of the GB trace in the image. Unfortunately the projected GB width has a minimum value over a tilt change of several degrees, and so the precision of this method is usually insufficient (Pumphrey, 1974).

The second type of method relies on tilting the foil in order to observe and measure changes in the image of the GB. Two vectors in the GB have to be selected and defined by measuring their coordinates relative to a reference frame. After tilting, the coordinates are remeasured and N can then be calculated (Young *et al*, 1973). The disadvantages of this method are that there is only limited availability of suitably defined vectors, such as GB triple points, and large tilts are required. Consequently this method is only suitable for a small proportion of GBs.

A similar method utilises a graphical construction or a linear regression whereby the projected GB width (which is measured directly from a micrograph) and the foil tilt angle (which is read from the goniometer) are obtained for at least three different specimen tilts (Marrouche *et al*, 1987). The following function of the projected GB width and tilt angle is used:

$$P/\cos \Omega_f = - W \sin \varphi' \tan \Omega_f + W \cos \varphi' \qquad (4.2)$$

where P is the projected GB width, W is the true GB width, Ω_f is the foil tilt angle (which is zero in figure 4.5a because the foil is horizontal) and φ' is the GB inclination angle measured from the *foil surface* to the GB plane. (Note that φ' is the complement of φ since φ is the GB inclination angle measured from the *foil normal* to the GB plane). Both W and φ' are obtained from the gradient and intercept, $W\sin\varphi'$ and $W\cos\varphi'$ respectively, of a linear regression of equation 4.2. Figure 4.5b shows the regression analysis for a typical set of experimental data obtained using a double tilt holder. Disadvantages of this method are that large tilts are necessary in order to effect measurably large changes in the projected GB width, and additional errors may be introduced from inaccurate calibration of the goniometer stage.

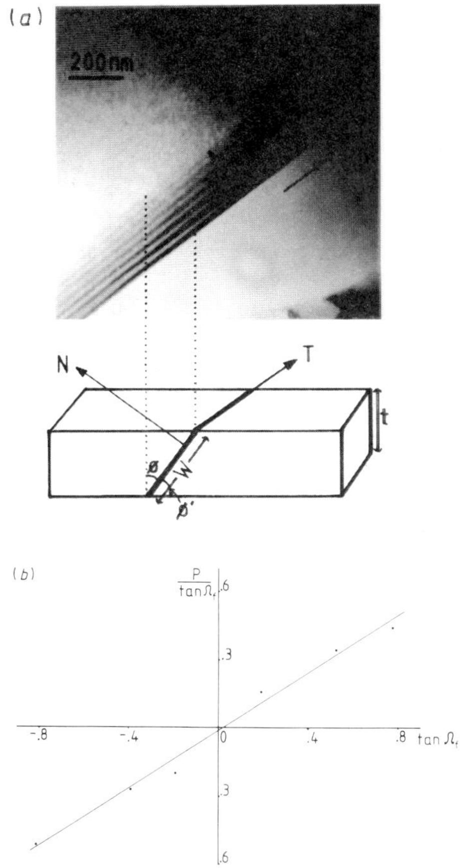

Figure 4.5 (a) Micrograph of an inclined GB in a TEM thin foil illustrating the recognition of the top and bottom of the foil from the contrast of the thickness fringes in the image of the projected GB, plus a schematic diagram of the parameters *P, T, W, t,* φ, and φ' which are defined in the text. These parameters are needed to obtain *N*, the GB plane normal (see also figure 5.7). (b) Illustration of the regression analysis for obtaining φ' using equation 4.2 (Experimental data supplied courtesy of M. Dechamps).

If both the thickness of the foil at the GB and the projected GB width are known, then φ can be calculated very simply from (Randle, 1989):

$$\tan \varphi = P/t \tag{4.3}$$

where t is the foil thickness as shown in figure 4.5a. The sense of φ must also be known, i.e. if it is a clockwise or anticlockwise rotation from the beam direction. In other words we need to know which edge of the GB projection intersects the top of the foil as it is viewed in the TEM. The top of the foil is evident from the image because the intensity fringes in the GB are stronger near the intersection with the top of the foil and are weaker near the lower intersection, as illustrated on figure 4.5a.

Techniques based on both convergent beam electron diffraction (CBED) and electron energy loss spectroscopy (EELS) can be used to measure foil thickness to an accuracy of 5% (Scott and Love, 1987). The EELS technique is only applicable for foil thickness up to approximately 100nm at a typical accelerating voltage of 120kV (Egerton, 1981). The GB inclination angle can be determined more reliably where foils are in the thickness range 150-250nm because for these cases large projected GB widths are achievable and thus the error on measuring P is minimised. CBED is therefore the better choice for foil thickness measurements in the present context since it can be used to measure thicknesses in the range 50-250nm when the accelerating voltage is 120kV.

The measurement of foil thickness by CBED is well-documented elsewhere (Allen and Hall, 1982; Ecob, 1985) and so will be outlined only briefly here. The requirement is to obtain a two-beam CBED pattern which does not have excessive interference from other reflections but also has sufficient intensity in the discs. 220 has been shown to be a suitable reflection. Figure 4.6 shows an example of a typical two-beam CBED pattern in nickel. The ratios of the angular separations of the intensity maxima and minima which can be seen in the dark field disc to the relevant Bragg angle are then used for a linear regression analysis of the following relationship:

$$(s_i)^2 = -(1/E_g)^2(1/n_k)^2 + (1/t)^2 \qquad (4.4)$$

where s_i are the values of the deviation parameter at the positions of intensity extrema in the dark field disc of the CBED pattern, n_k are integers and E_g is the extinction distance. In practice, s_i are obtained from the distance between the zero position in the dark field disc (measured off an enlarged print) and each extrema divided by the separation distance between the centres of the bright field and dark field discs. The foil thickness is obtained from the intercept in the linear regression analysis of this function. Once the relevant input data have been measured off an enlarged print of the diffraction pattern, the fit of the best straight line for various values of n_k is found rapidly by a computer routine.

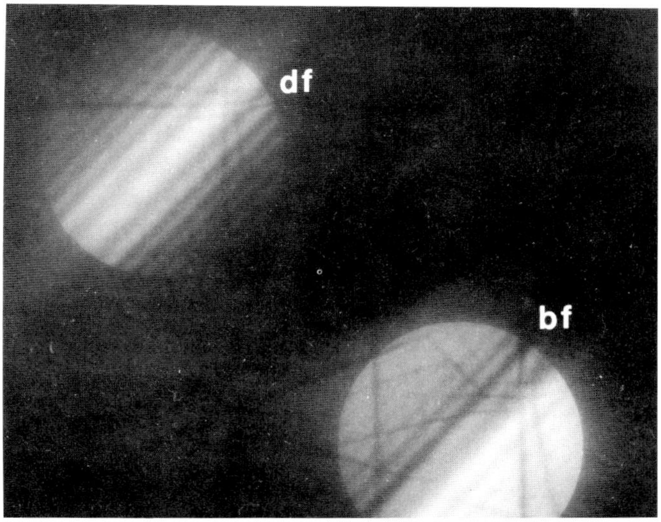

Figure 4.6 Two-beam 220 CBED pattern from an fcc steel showing the intensity extrema in the dark field (df) disc.

The amount of tilt required to reach a 220 reflection is usually only a few degrees because these Kikuchi lines are abundant in a diffraction pattern from an fcc material. It is important to minimise the amount of tilt so that for small values the measured thickness can be considered to be the true foil thickness. If there is not a 220 reflection conveniently available within a few degrees of tilting, a 311 reflection can be used. Of course, t cannot be measured at the GB position precisely; it is measured close to it.

Once the foil thickness, and hence φ, is known, the only other input data which is required for computation of N is a diffraction pattern from each interfacing grain and the crystallographic direction of the GB trace, $\boldsymbol{T}$, which is marked on figure 4.5a. The trace direction is obtained from the diffraction pattern of one of the grains, taking into account the rotation of the image with respect to the diffraction pattern for that particular microscope. The image/diffraction pattern rotation is caused by the spiral trajectory of the electrons as they pass down the column of the microscope. The calibration only needs to be performed once for a particular instrument and this is usually done with a specimen which is known to exhibit a well-defined crystallographic direction, such as molybdenum trioxide crystals which have edges which are precisely parallel to 001. Table 4.2 is a typical calibration for combinations of magnification and camera length.

TABLE 4.2

RELATIVE ROTATION BETWEEN IMAGE AND DIFFRACTION PATTERN - SPECIMEN VALUES FOR ONE MICROSCOPE

Camera length 800mm:

(Negative sign indicates anticlockwise rotation)

mag.x10^3	rotation°	mag.x10^3	rotation°
10	42	36	-59
13	44	46	-56
17	42	60	-54
22	37	80	-50
28	51	100	-43

Additional rotations for other camera lengths:

CL mm	rotation°	CL mm	rotation°
210	30	1150	-17
290	23	1600	-35
400	15	2300	-21
575	8		
800	0		

In summary, the following parameters must be measured in order to obtain *N* by the 'foil thickness' method, which is considered to be the best TEM-based choice: the foil thickness as measured from a two-beam CBED pattern, the direction of the boundary trace, the projected GB width *P* and the orientation of each interfacing grain, obtained from a single diffraction pattern from each. The procedure for calculating *N* from these parameters is described in section 5.5.

4.4 ELECTRON BACK-SCATTER DIFFRACTION IN THE SEM

The essential requirement in the formation of a Kikuchi pattern is that the electron beam impinges on all lattice planes at the Bragg angle, as explained in section 4.2. One way in which this criterion can be met in the SEM is by altering the beam scanning system so that the beam rocks about a point on the specimen surface, thus sampling all the Bragg angles. This is the principle of the SAC technique (Joy, 1975; Dingley, 1981; Lorenz and Hougardy, 1988). In order to be effective SAC requires a flat, polished specimen surface and the sample area is about 10μm.

A second way in which diffraction occurs in the SEM is if the specimen is tilted to make a small angle with the incident beam as in figure 4.7. This configuration allows the path length of the electron beam in the crystal to be considerably shortened with the result that most of the electron beam is diffracted out of the specimen rather than absorbed (Venables and Harland, 1973; Dingley, 1991). By contrast if the specimen is viewed flat in the SEM for conventional imaging very little of the electron beam is diffracted out. The diffraction effect obtained from the simple step of tilting the specimen towards the incident beam is called electron back-scatter diffraction, which is also sometimes referred to as back-scatter(ed) Kikuchi diffraction, BKD. EBSD is now considered to be the principal diffraction technique in the SEM for orientation work (see Table 4.1).

4.4.1 General description of EBSD

Electron back-scatter diffraction was developed as a practical experimental tool in the early 1980s (Dingley, 1981) and has now become the most important technique for measuring individual orientations. In turn it is therefore also a major technique for obtaining misorientations and other GB parameters (Randle and Ralph, 1988e,f,g; Randle and Furley, 1991; Randle, 1992a, 1992b). Because of the importance of EBSD in the context of GB geometry, and also because it is still a relatively new technique, a general description of EBSD and its operation is included here. A detailed description of the practical aspects of EBSD and microtexture determination can be found elsewhere (Randle, 1992c).

The EBSD patterns, generated by the interaction of the primary beam with the tilted specimen, are captured via a phosphor screen interfaced to a low-light TV camera and control unit. The Kikuchi patterns are viewed on a monitor screen and interrogated in real time. The components of an EBSD system are a tilted specimen holder, phosphor screen, low light TV camera, camera control unit, TV monitor and microcomputer. Some image enhancement/frame store capability is also usual. Typical operating conditions are 20-30kV accelerating voltage and probe current 0.1-50nA. The best resolution is 100 - 200nm surface diameter, which is achieved by operating in the low range of probe current and accelerating voltage. However, these operating conditions also reduce the signal-to-noise ratio and for practical purposes an image enhancement facility must be used to improve the pattern quality. Because the specimen is highly tilted with respect to the primary beam, the specimen/beam interaction depth is about 20nm, which is greatly reduced compared to the interaction depth, about 1μm, of the beam with a flat specimen.

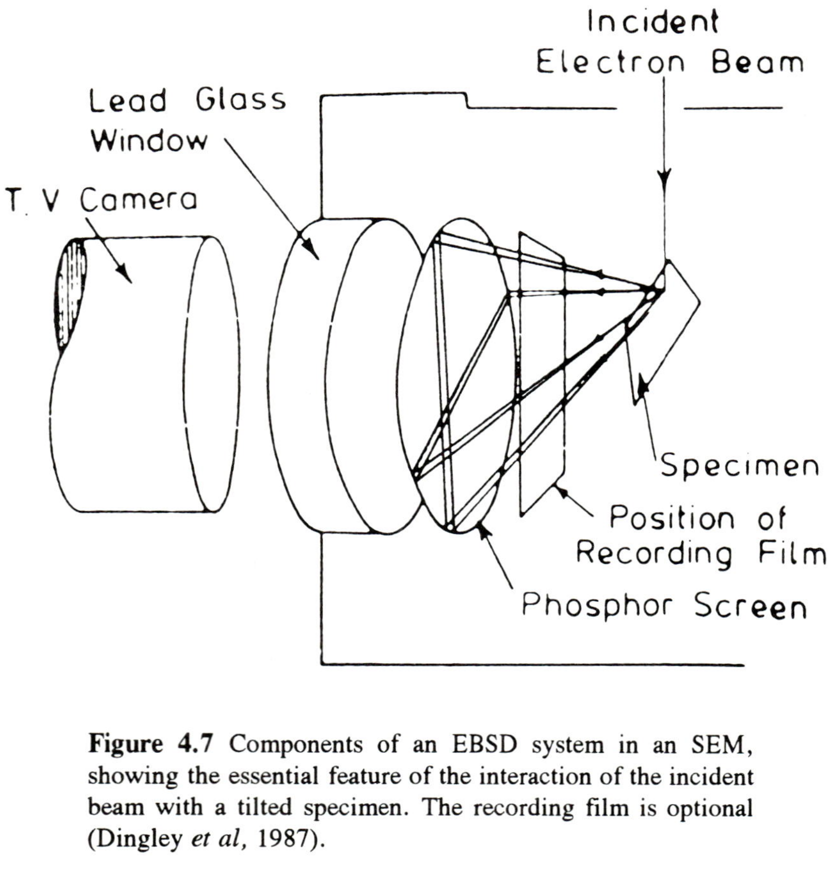

Figure 4.7 Components of an EBSD system in an SEM, showing the essential feature of the interaction of the incident beam with a tilted specimen. The recording film is optional (Dingley *et al,* 1987).

The Kikuchi patterns shown in figure 4.2 were obtained using EBSD. For these cases the patterns were recorded directly, that is, by inserting photographic film into the microscope column. Figure 4.8 is a comparable pattern taken from the central portion of the circular phosphor screen as viewed on the monitor screen after some image enhancement. The direct recording captures a very large angular range in the diffraction pattern, approximately 50° x 80°, whereas the monitor pattern still covers a large angular range but somewhat less than the direct photograph. The angular range of both these EBSD patterns is far greater than that of either TEM or SAC patterns, which facilitates pattern recognition. Furthermore, the directly recorded patterns show more detail than the monitor pattern. This point, coupled with the larger range, is important if EBSD is being used as a tool for phase indentification, but not for routine orientation measurement in cubic materials where it is sufficient to view patterns on the monitor.

Turning now to specimen preparation, the primary requirement for preparing specimens which are suitable for EBSD work is that they should be relatively undeformed in approximately the top 20nm of the specimen surface, which is a typical interaction depth of the electron beam with a steeply inclined specimen. The usual specimen preparation route is metallographic grinding and diamond polishing followed by etching; the etching performs the dual function of removing surface deformation accrued during the mechanical preparation stages, and revealing the microstructure. For GB work a preferred etchant is one which reveals the positions of the GBs; otherwise imaging in the backscattered mode can provide grain contrast (Lorenz and Hougardy, 1988) or the position of GBs is inferred from the change in diffraction pattern.

A certain amount of damage can be tolerated within the interaction depth because although the diffraction pattern becomes more diffuse, i.e. the Kikuchi lines broaden, the position of zone axes remains the same (Randle and Dingley, 1990b). Furthermore, the deformation is often inhomogeneously distributed, e.g. as a polygonised structure, and in these cases the probe can be sited in regions of least lattice strain. Thus it is not possible to quantify a level of deformation at which EBSD become invisible because this will be influenced by the dislocation arrangement. The degree of pattern diffuseness can be used as a qualitative or semi-quantitative guide to the level of plastic strain in the material. More details can be found elsewhere (Dingley and Randle, 1992).

4.4.2 Grain misorientations

Once the specimen has been prepared, mounted on the special tilted holder and an image formed of a suitable area for study, the steps taken to obtain a diffraction pattern and thus an orientation measurement are analogous to those for microdiffraction in TEM. A small probe is sited on a selected grain by switching to 'spot' mode and the diffraction pattern is then viewed on the monitor screen. For orientation measurements using EBSD it is not usually necessary to record a pattern since the measurements necessary to compute the orientation are inputted on-line from the real-time pattern.

Indexing of EBSDs on the monitor screen is an interactive process between the operator and software routine. The most commonly available EBSD software at the time of writing requires the operator to identify two zone axes in the pattern and input their coordinates with a cursor. Other software does not require recognition of zone axes in the pattern but instead

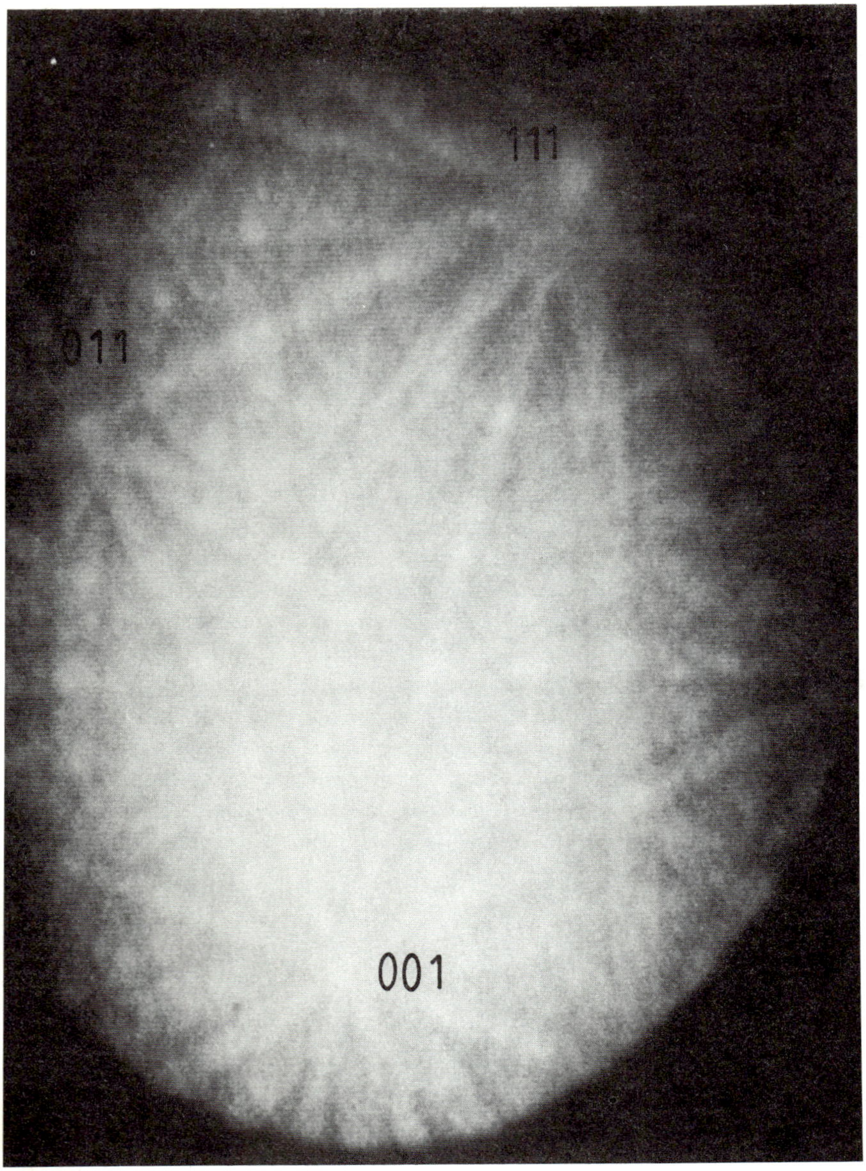

Figure 4.8 EBSD pattern from tungsten recorded off the monitor screen (Courtesy J.A. Venables).

requires at least three unidentified Kikuchi lines to be picked out with the cursor whereupon a look-up table stored in the memory is used to orient the pattern correctly. Automatic pattern orientation routines which require no operator input are also under development (Schmidt *et al,* 1991; Juul Jensen and Schmidt, 1991; Wright and Adams, 1991; Wright *et al,* 1991). However, the automatic route is not in its present form suitable for GB studies because for such investigations the operator needs to site the probe correctly with respect to the local environment and to oversee the correlation of the diffraction patterns with the image.

Obtaining diffraction data using EBSD requires a very specific calibration routine, which is in contrast to the TEM route where the only calibration required is that for general microscope operation. The reason for this difference is that TEM diffraction patterns (and also SAC patterns) are concentric with the primary electron beam. This means that the origin of the projection plane, the pattern centre, coincides with the physical centre of the pattern. For EBSD the pattern centre does not coincide with the physical mid-point of the pattern; rather it is governed by the position of the pattern source point on the specimen surface. This is defined as the point at which the primary electron beam impinges upon the specimen. Hence the pattern source point can be thought of as the centre of the reference sphere of the gnomonic projection in figure 4.3, and the pattern centre is the point *N*. In order to convert the distances on the EBSD pattern to angles, it is also necessary to know the radius of the reference sphere, i.e. the distance from the pattern source point to the pattern centre, *L*. The geometry of the EBSD system with respect to these parameters is shown in figure 4.9. The EBSD calibration routine therefore has to ascertain the coordinates of pattern centre and the distance *L* for a specimen at a fixed tilt angle within the microscope. The X-direction of the reference axes is taken to be horizontal in the microscope, i.e. parallel to the transverse control of the specimen stage movement, and the Y-direction is therefore parallel to the 'back-front' movement control as shown in figure 4.9.

Accurate calibration of the system has proved to be the major technical challenge in implementing EBSD to measure orientations with acceptable accuracy. The conventional calibration method at present utilises a silicon single crystal, cleaved to show (001), and a purpose-built holder, accurately inclined at 70.5° to the horizontal. This geometry means that the 114 direction from the silicon crystal is horizontal, because the angle between [001] and [114] is 19.5° which is the complement of the inclination angle of the holder, as illustrated on figure 4.10a. There are various refinements to this method and also other methods which do not utilise a special crystal. These are all documented elsewhere (Randle, 1992c; Dingley and Randle, 1992). One alternative method to the standard silicon calibration involves moving the phosphor screen between two positions and recording the coordinates relative to the screen of at least three poles for each of the two positions. The pattern centre (beam normal, BN) is then defined by the intersection of three lines joining the 'before' and 'after' positions of the poles as shown in figure 4.10b.

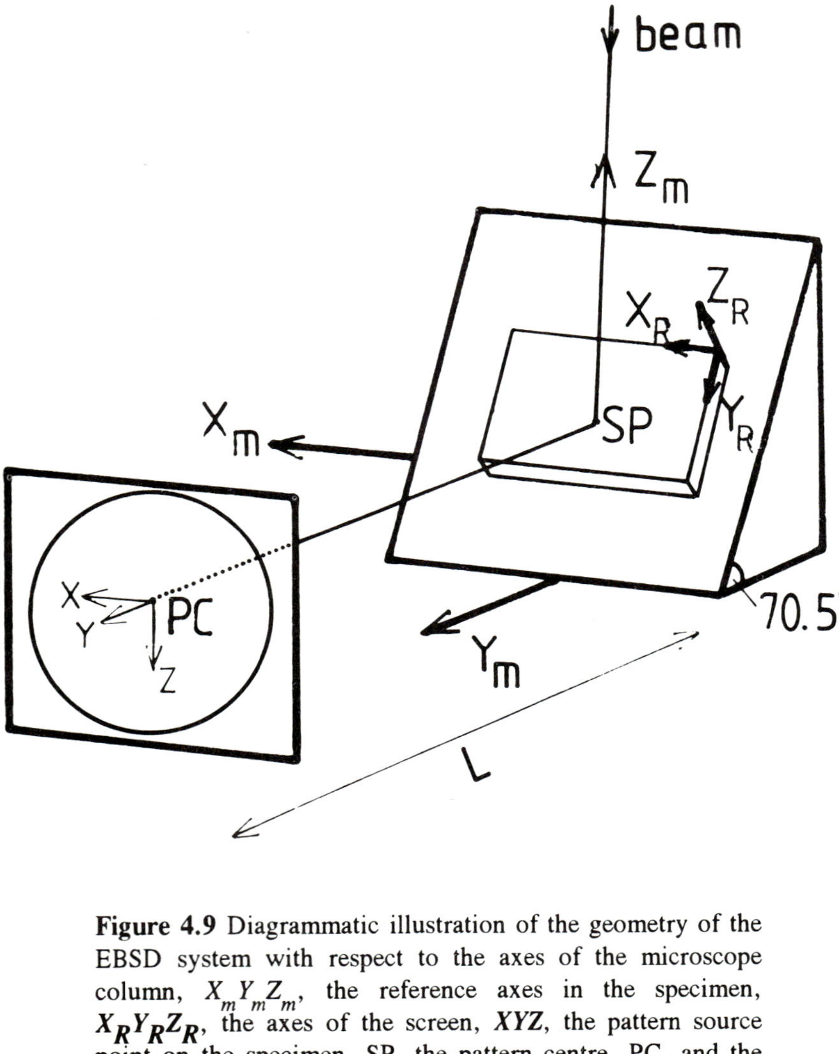

Figure 4.9 Diagrammatic illustration of the geometry of the EBSD system with respect to the axes of the microscope column, $X_mY_mZ_m$, the reference axes in the specimen, $\boldsymbol{X_RY_RZ_R}$, the axes of the screen, ***XYZ***, the pattern source point on the specimen, SP, the pattern centre, PC, and the distance between SP and PC, *L* (Randle, 1992a,c).

The important point with regard to using EBSD for GB studies is that accuracy is of paramount importance. With care, a misorientation can be measured to a precision of 0.5-1.0°. However, inaccuracy in the calibration routine could degrade this precision by up to 3°, which is not acceptable if the data are subsequently to be used to identify CSL GBs (see section 5.4) or other geometrically special GBs. There are several factors which can affect the

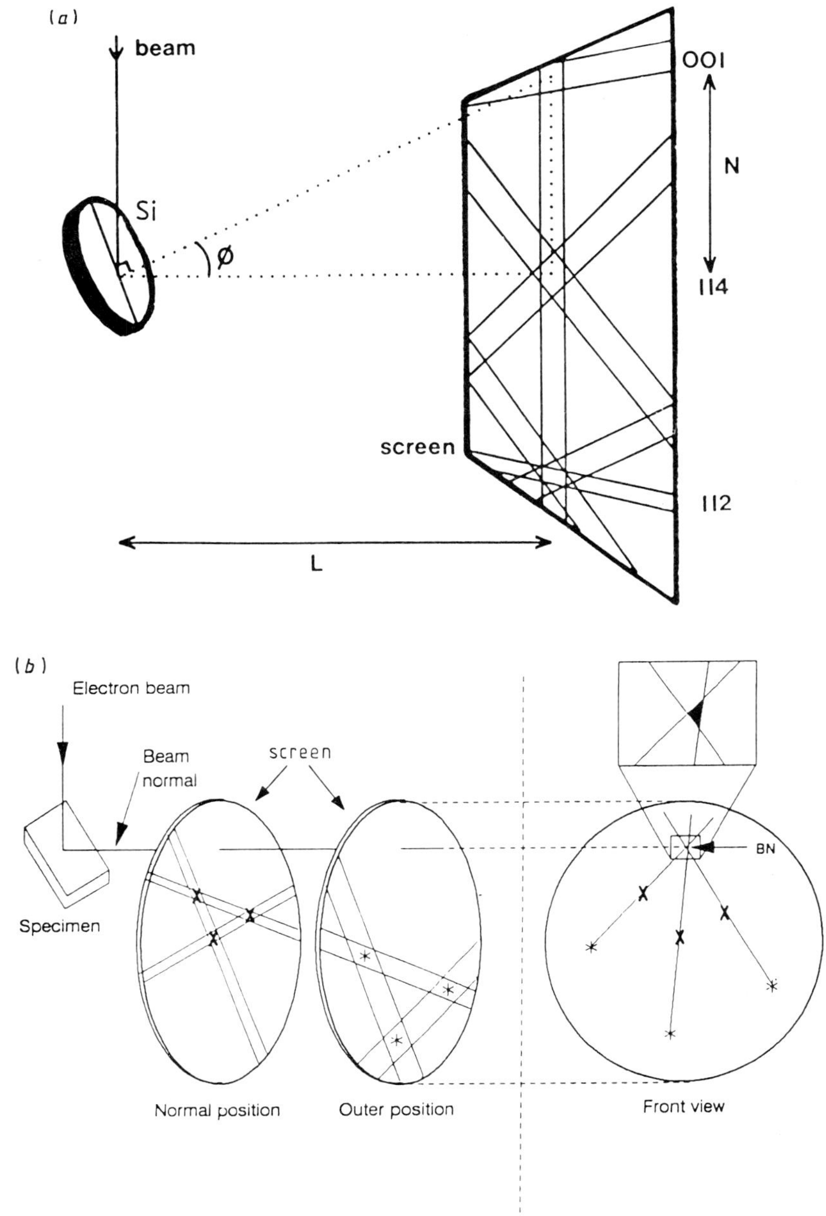

Figure 4.10 Calibration of the EBSD system. (a) Established calibration method using an 001 silicon crystal tilted at 19.5° to the beam. (b) Alternative calibration method involving moving the phosphor screen and recording the positions of the same three zone axes for both screen positions (Courtesy of J. Hjelen) (Hjelen, 1992).

accuracy. If the silicon calibration method is used, the position of the calibration crystal and the specimen must be invariant with respect to the microscope geometry, which includes having the surface of both parallel to the inclined holder, i.e. 70.5° to the horizontal. Another consideration is the specimen height. Both the calibration and the orientation measurement routines need the specimen height as an input parameter because this will influence both the position of pattern source point and pattern centre and the distance from the pattern source point to the pattern centre. For most microscopes the specimen height is only known to the nearest millimetre, as displayed on the microscope column, which could feed through as an error of up to 1.5%. Consequently for GB studies it is advisable to keep the working height constant between calibration and data collection, which in practice can be achieved readily.

If the GB parameters are to be related to the specimen geometry, i.e. if the three 'extrinsic' degrees of freedom are to be accessed, then it is important that the specimen is positioned accurately with respect to the specimen holder and hence to the microscope axes. Because of the bulk nature of specimens for EBSD, specification of external axes can be done with greater ease and accuracy (i.e. 1-2°) than is the case for TEM thin foils.

4.4.3 Grain boundary planes

The experimental requirement for measuring the orientation of GB planes in bulk polycrystals in the SEM is to obtain the inclination angle, φ, in addition to the orientation of both grains. We will consider two methods here.

The first method is based on a simple serial sectioning procedure (Randle and Dingley, 1992). Hardness indents can be used as a guide to the depth of specimen which has been removed by grinding, because the surface angle (i.e. the angle between two opposing faces) of the diamond indenter is known and accurate to better than 1°. Figure 4.11a and b show an indent before and after grinding 50μm off the specimen. The amount of GB displacement on the specimen surface can be measured (figure 4.11c) and thus φ can be found from a knowledge of the GB displacement over a specific depth. The precision of this method for obtaining φ is degraded if the depth of material removed becomes too small. The smallest depth change acceptable is about 20μm which restricts the grain size of the material for which this method can be applied to greater than this amount.

Another measurement method for obtaining the GB plane normal is a 'two-surface trace analysis' approach, which was first carried out for investigations using X-rays (Andrejeva, 1978). A GB trace is observed on two inclined (preferably perpendicular) and adjoining surfaces of a specimen as illustrated schematically in figure 4.12a. The specimen therefore has to be metallographically prepared on the two adjoining faces, taking care to retain the 'sharpness' of the edge as shown in figure 4.12b. In figure 4.12b the specimen has been tilted in the SEM so that both faces can be viewed together.

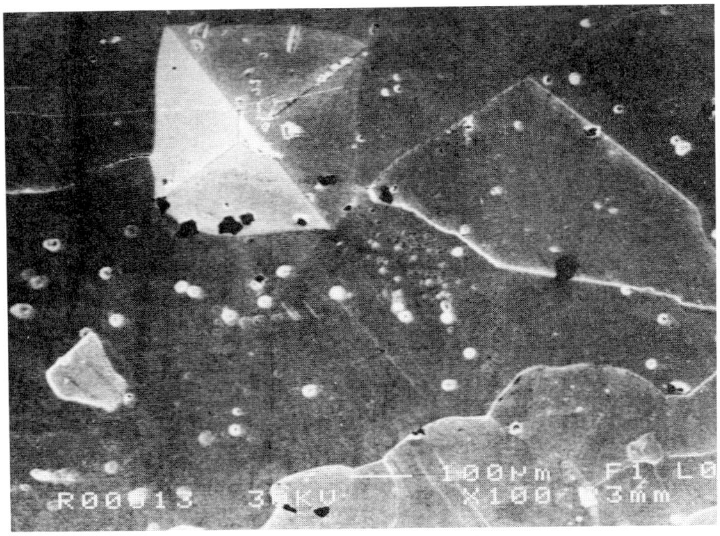

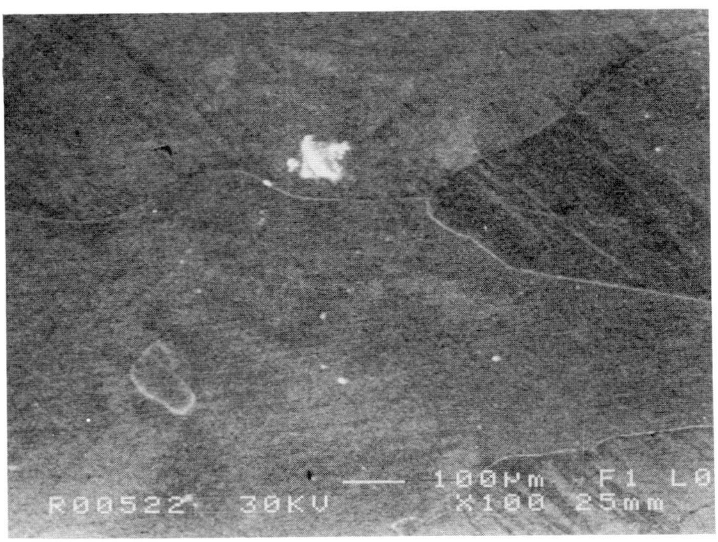

Figure 4.11 A method for obtaining the GB inclination, φ, in the SEM. (a) and (b) are micrographs of hardness indent in an iron specimen (a) before and (b) after grinding 50μm off the specimen. In (c) the positions of the GBs before and after the sectioning are shown.

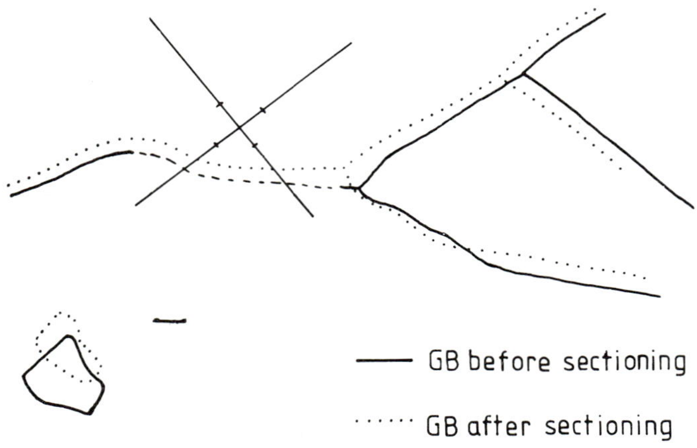

Figure 4.11 (*continued*)

The specimen is mounted in the microscope so that the common edge between the two surfaces is accurately parallel to the X-axis of the pre-inclined specimen holder. If the GB trace makes an angle α with the X-axis on one face of the specimen (UF, the 'upper face') and an angle β with the 'lower face', LF, then α and β define two points on the GB plane as shown on the stereogram in figure 4.12c (Randle and Dingley, 1989, 1990a).

The input data for this analysis for N are therefore α and β, which are measured off micrographs of the tilted specimen, and the orientation of each grain. α and β measured from a micrograph of a tilted specimen must be corrected to allow for the distortion in the image introduced by the tilt. The specimen tilt angle is usually 70.5°, the angle of the pre-inclined tilt holder, so that EBSD data can be collected concurrently with the trace mapping. The relationship between the true α and β and the projected angles α' and β' is shown on figure 4.12d, and the true angles are calculated thus:

$$\tan \alpha = \tan \alpha' / \cos \Theta \qquad (4.5a)$$

$$\tan \beta = \tan \beta' / \cos(90° - \Theta) \qquad (4.5b)$$

where Θ is 70.5° for the conventional EBSD configuration. Small grain sizes may degrade the accuracy in measurement of α and β because the GB trace lengths become short. However, grain sizes down to about 10μm are acceptable for this method, which is a lower limit than for the serial sectioning method. Both methods necessarily rely on the assumption that the GB is planar rather than curved over the measurement distance.

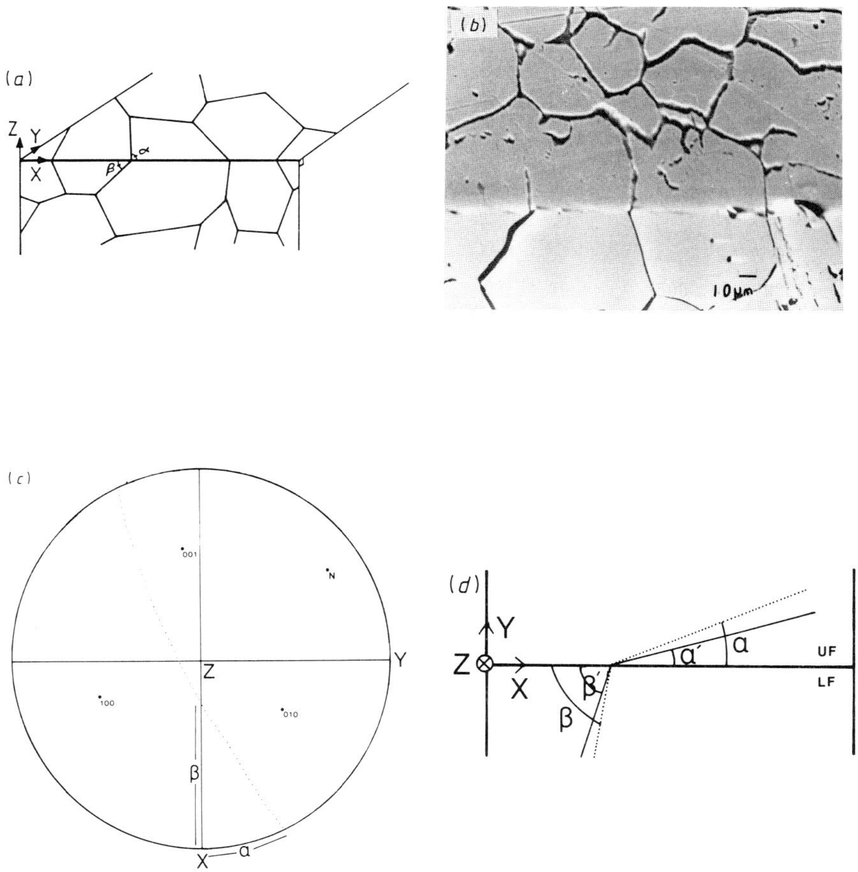

Figure 4.12 A method for obtaining the GB plane inclination, φ, in the SEM. (a) Schematic illustration of a specimen prepared on two mutually perpendicular and adjoining surfaces showing the angles α and β that a GB makes with the X-axis. (b) Nickel specimen prepared as in (a) and viewed looking down onto the X-axis. (c) Stereographic interpretation of the procedure for obtaining the GB plane normal, *N*, from the information shown in (a) (see also section 5.5). (d) Relationship between the true angles that the GB trace makes with the X-axis and those viewed and measured from a tilted specimen (Randle and Dingley, 1989, 1990).

5

DATA INTERPRETATION AND PROCESSING

5.1 INTRODUCTION

The data processing part of GB geometrical analysis involves the analysis of Kikuchi diffraction pattern pairs to calculate the misorientation matrix, output it in a convenient form (e.g. in the angle/axis notation) and interpret it in terms of any geometrically special relationships (e.g. CSLs). If the appropriate spatial measurements have been made the data can also be processed to extract the GB plane normal, which then allows a more detailed analysis and interpretation of the GB geometry. This chapter will describe the relevant procedures for calculating a misorientation matrix and GB plane normal from experimental measurements.

5.2 GENERAL PROCEDURE FOR CALCULATING A MISORIENTATION

Several software packages are available to process diffraction patterns and output a misorientation. Commercially available EBSD systems include dedicated software, and similar programs intended primarily for TEM can also be obtained (Heilmann *et al,* 1982). The primary requirement of diffraction pattern interrogation is the input of coordinates of at least two poles or Kikuchi line pairs. The Miller indices of the poles/lines can either be input by the operator or identified by a computer routine which compares the positions, widths and intersections of the marked Kikuchi line pairs with a stored look-up table of all possible configurations and selects the best pattern match. The number of possible options is reduced by inputting the coordinates of at least three and preferably more Kikuchi line pairs.

To obtain a misorientation the objective is to measure the orientation of two neighbouring grains relative to the same reference system, and so firstly we need to define a common reference system. The origin of the reference system of orthogonal axes $X_R Y_R Z_R$ is at the pattern centre. In the TEM X_R and Y_R can be chosen parallel to the sides of the plate and Z_R antiparallel to the beam. For EBSD X_R and Y_R are parallel to the left-right and forward-back traverses in the microscope respectively, and Z_R is antiparallel to the beam (see figure 4.9).

The objective of the diffraction pattern analysis is to express three orthogonal directions in the specimen in terms of their crystal coordinates, which is the definition of the orientation matrix (Bunge, 1985, 1987). Although specific procedures may vary in detail, the general procedure for determining an orientation from a diffraction pattern is to define three zone axes within the pattern in terms of both their specimen and crystal coordinates. This provides the necessary link between the specimen and crystal systems. The orientation is obtained by a matrix product involving the following three coordinate systems (Schmidt *et al,* 1991; Wright *et al,* 1991):

orientation (crystal/specimen transformation)	=	specimen/pattern transformation	x	pattern/crystal transformation	(5.1)

where the matrix expressing the transformation between the pattern and the crystal axes incorporates the configuration of the system apparatus, i.e. the position in the microscope of the viewing/recording screen relative to the specimen. For some EBSD systems this may be a 19.5° rotation between the screen and specimen surface (see sections 4.4.1. and 4.4.2). The 'crystal system' is the principal crystallographic directions, i.e. 100, 010 and 001, and the 'pattern system' refers to coordinates chosen in the Kikuchi pattern as viewed on the screen or photograph therefore is sometimes referred to as the 'screen system or coordinates'. It is also important to realise that the terms 'specimen system or axes' and 'reference system or axes' are synonymous since the reference axes are chosen to correspond to the specimen geometry, e.g. Z_R is normal to the specimen surface. Either nomenclature may be used in other texts. Here, for clarity, we will refer to 'specimen-reference axes'.

Figure 5.1 shows the relationship between the specimen-reference and pattern systems viewed in the 'TEM sense', that is, with the specimen and screen vertically juxtaposed so that the projection plane of the diffraction pattern is parallel to the specimen surface when the tilt goniometers are set to zero (Young *et al,* 1973). The corresponding diagram in the 'EBSD sense', which was illustrated in figure 4.9, has the screen vertical rather than horizontal. Furthermore, for EBSD the specimen-reference axes are defined by fixed directions in the microscope, therefore they project as a horizontal and vertical direction in the pattern, with Z_R normal to the pattern centre. For TEM, however, vectors are chosen in the pattern ($X'_P Y'_P Z'_P$ on figure 5.1) and then rotated parallel to the specimen-reference normal to the specimen surface. The procedures are described in the next section.

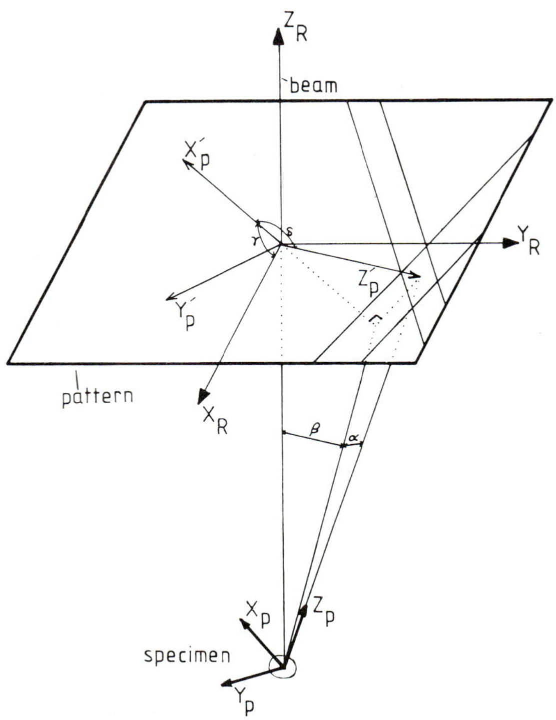

Figure 5.1 TEM configuration of the specimen, diffraction pattern projected onto the recording screen and beam direction. $X'_pY'_pZ'_p$ correspond to the projection in the diffraction pattern of the chosen vectors $X_PY_PZ_P$ in the specimen, which are in turn rotated to $X_RY_RZ_R$. The angles δ, γ, α and ß are described in section 5.3.

Three orthogonal directions, $X_PY_PZ_P$ in the specimen and their projections $X'_PY'_PZ'_P$ in the diffraction pattern are obtained from measurements off the pattern, either by hand off an enlarged print or by directing a computer cursor or mouse, then passed to a computer routine. The x and y coordinates of vectors which define points in the pattern are measured relative to the directions X_R, Y_R and the pattern centre. The z coordinate is the specimen to screen distance, L, which in TEM is the camera length. L can be obtained by measuring the projected distance D in the diffraction pattern between the pattern centre and another pole and using the relationship

$$L = D/\tan\Theta_p \tag{5.2}$$

where Θ_p is the angle between the pattern centre and the pole. For EBSD this determination is either carried out using a silicon calibration crystal cleaved along (001) (which means that the pattern centre is 001 if the screen and specimen surface are parallel or 114 if the screen and specimen surface are inclined at 19.5° to one another) or by an alternative method (see section 4.4.2). For TEM either a single crystal is used to obtain an on-zone diffraction pattern or a polycrystalline specimen is searched to find a grain which is either on-zone or requires only a very small tilt to align the beam with the zone axis. Larger tilts are not permissible because the calculations assume that the screen is parallel to the specimen surface. Figure 5.2 shows a Kikuchi pattern from nickel obtained using TEM where the beam direction coincides with a zone axis, for this case 114. The second pole is 113 and the angle c between 114 and 113 can be calculated using the following well known expression

$$\cos c = \frac{u_1u_2 + v_1v_2 + w_1w_2}{((u_1^2+v_1^2+w_1^2) \times (u_2^2+v_2^2+w_2^2))^{1/2}} \tag{5.3}$$

where $u_1v_1w_1$ and $u_2v_2w_2$ are two poles and c is the angle between them. For this case L is 190mm.

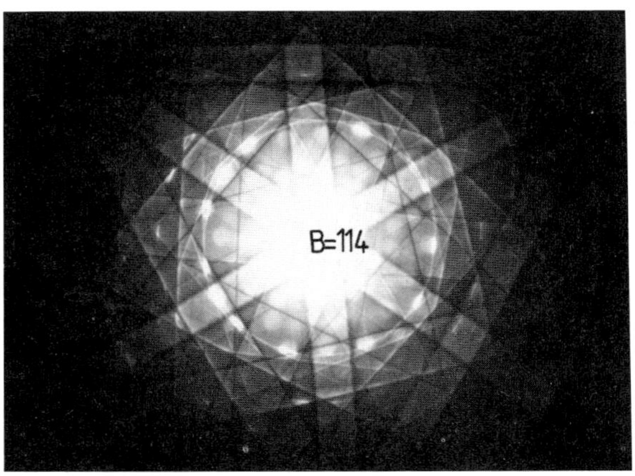

Figure 5.2 A diffraction pattern taken in the TEM where the beam direction coincides with 114.

The orientation of each grain is expressed as a 3 x 3 matrix which gives the direction of the three crystal axes with respect to the specimen-reference system. The first row of the matrix is given by the cosines of the angles between 100 and $\boldsymbol{X_R}$ (α_1), 100 and $\boldsymbol{Y_R}$ (β_1) and 100 and $\boldsymbol{Z_R}$ (γ_1), and similarly for the second and third rows with 010 and 001 respectively. The orientation matrix **A** is then given by

$$\mathbf{A} = \begin{bmatrix} \cos\alpha_1 & \cos\beta_1 & \cos\gamma_1 \\ \cos\alpha_2 & \cos\beta_2 & \cos\gamma_2 \\ \cos\alpha_3 & \cos\beta_3 & \cos\gamma_3 \end{bmatrix} \tag{5.4}$$

Thus, for example, the direction cosines of the crystal direction which is parallel to the specimen normal, $\boldsymbol{Z_R}$, are the cosines of $\gamma_1\gamma_1\gamma_3$. It is important to note that for most EBSD software the arrangement of the rows and columns of equation 5.4 are swapped. However, the standard, conventional form of an orientation matrix is as shown in equation 5.4 (Bunge 1985, 1987). Once the orientations of two neighbouring grains have been obtained relative to the *same* reference axes the misorientation matrix **M** is obtained from

$$\mathbf{M} = \mathbf{A}_2^{-1}\mathbf{A}_1 \tag{5.5a}$$

where $\mathbf{A}_1$ and $\mathbf{A}_2$ are the orientation matrices of two neighbouring grains. Equation 5.5a gives the misorientation relative to the specimen-reference axes, i.e. the way in which each orientation is measured experimentally. The inverse of equation 5.5a gives the misorientation, **M'** relative to the crystal coordinate system of grain 1 (Haessner *et al,* 1983):

$$\mathbf{M'} = \mathbf{A}_2\mathbf{A}_1^{-1} \tag{5.5b}$$

These are the basic principles for extracting a misorientation from two diffraction patterns, and in the next two sections we will examine a typical example. Once the misorientation matrix has been obtained, further calculations will be needed to carry out some or all of the following: finding the angle/axis pair, the lowest angle symmetry-related solution, Euler angle representation and CSL or CAD relationships. If the GB plane normal is known the GB can be analysed according to its tilt and twist components. All these procedures, and the relationship between them, are summarised in Table

5.1. The theoretical aspects of these procedures were described in chapters 2 and 3, and in the present chapter they will be illustrated further through experimental examples. Further computations again are required for representation and display of sample populations of GB data, and these are discussed in chapter 6.

TABLE 5.1

PROCEDURES FOR PROCESSING MISORIENTATION DATA

Procedure	Equations	Sections	Tables
Diffraction data → matrix	2.6, 5.1, 5.4, 5.5	2.3.1, 5.2	
Matrix → Θ/UVW	2.8 - 2.10, 5.6 - 5.12	5.3	
Θ/UVW → Θ_{min}/UVW	2.20, 5.10	2.4, 5.3	
Θ/UVW → CSL	3.8, 5.13 - 5.15	3.2, 5.4	3.1, 3.3, 3.4
Θ/UVW → CAD	3.8, 3.10	3.3.1	3.7
Matrix, α, φ → GB plane	2.22 - 2.26 5.15	2.2, 2.7, 5.5	
GB plane → classification	2.27, 3.10	3.2.4, 5.5	3.5, 3.6
Matrix → Euler angles	2.14-2.18	2.3.2, 2.7,	
Euler angles → CSL		3.2.5	6.3, 6.4

5.3 DIFFRACTION PATTERN ANALYSIS TO OBTAIN A MISORIENTATION

The most commonly used procedures for diffraction pattern interpretation are based on the so-called 'pole and line' method (see section 5.3.1). This means that the required input data include a selected zone axis and a Kikuchi line. There are at least two variations on this method. For one variation the selected pole is any convenient, usually low-index, zone axis and the Kikuchi line is in that zone. Thus the zone axis, the direction perpendicular to the Kikuchi line and the cross product of these two form a pattern system set of axes. The transformation between the pattern system and the specimen-reference system is performed via three sequential Euler-type angles which are obtained from the pattern by measurement and calculation.

In a second variation on the 'pole and line' method the selected pole is the beam direction, ***B***. Unless ***B*** is exactly on zone its direction cosines must be calculated by 'triangulation' (see subsection 5.3.2) which involves a knowledge of two or three zone axes plus the angles between each of them and ***B***. A Kikuchi line is selected in the pattern and a pattern system set of axes is set up. Because for the triangulation method the pattern centre already coincides with the *Z*-axis in the pattern system, only one rotation is needed, rather than three, to rotate the pattern system onto the specimen-reference system.

The two variations on the pole and line method outlined above require similar amounts of operator input and computation. Each method will now be considered in more detail through the example of the diffraction patterns shown in figure 5.3a and b, which are two patterns from neighbouring grains in an austenitic steel specimen. In subsection 5.3.3 we also illustrate the use of the stereogram for obtaining the misorientation between the grains in figures 5.3a and b. These patterns were obtained using TEM. A procedure based on the same principles is used for obtaining a misorientation from an EBSD patterns.

5.3.1 'Pole and line' method

The location of the main Kikuchi bands in figures 5.3a and b are reproduced diagrammatically in figures 5.4a and b and relevant measurements for the analysis are marked. Hereafter these will be referred to as grains 1 and 2. In grain 1 the *z* and *x* vectors in the pattern system are 112 and $\bar{2}20$ respectively, and for grain 2 they are 114 and $2\bar{2}0$. Kikuchi lines may be indexed from first principles using Bragg's Law to identify the 'family' of planes from the relative band widths and a stereogram to identify the specific indices of the plane, or more conveniently reference can be made to an indexed crystallographic 'map' (Loretto, 1984; Randle, 1992c). Thus for each grain a matrix **P** can be formulated which represents the transformation between the crystal system and the pattern system. In other words the 1st, 2nd and 3rd rows of **P** define respectively the *x*, *y* and *z* axes of the pattern system $(\boldsymbol{X'_P Y'_P Z'_P})$ in crystal coordinates.

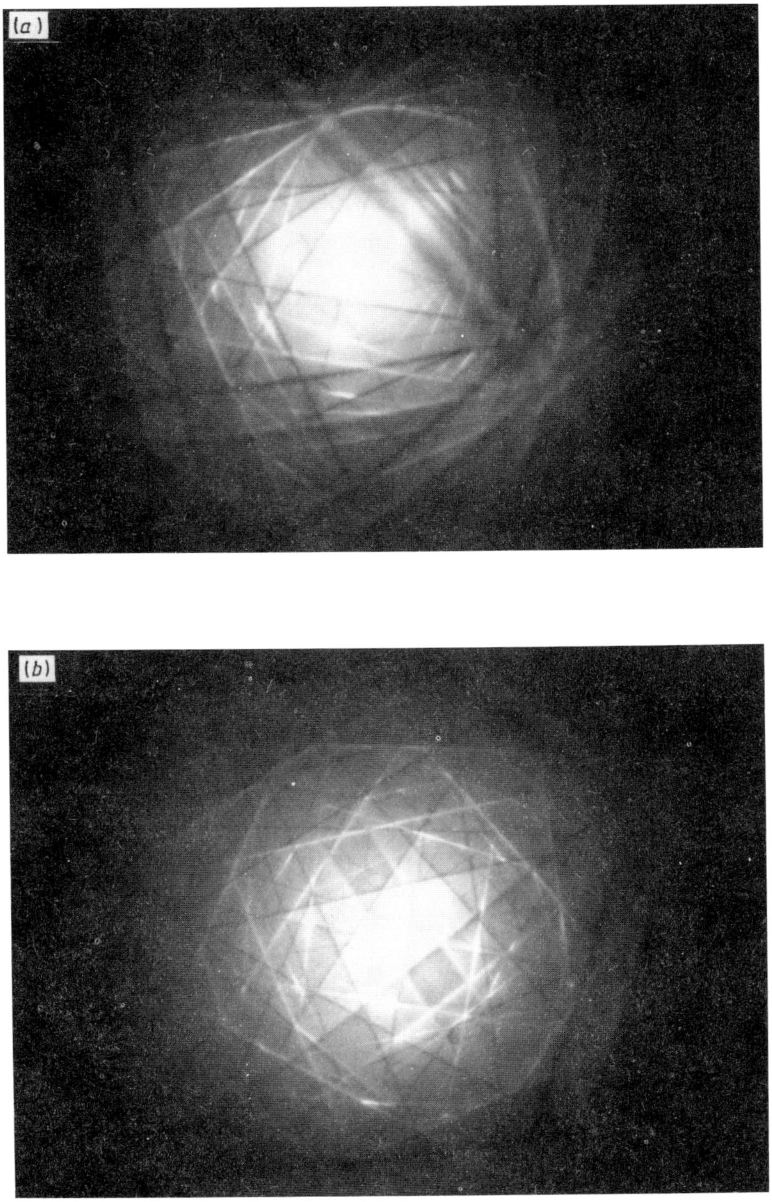

Figure 5.3 Diffraction patterns from neighbouring grains in an austenitic steel, used in this chapter to illustrate procedures for calculating the GB misorientation (see also figure 5.4).

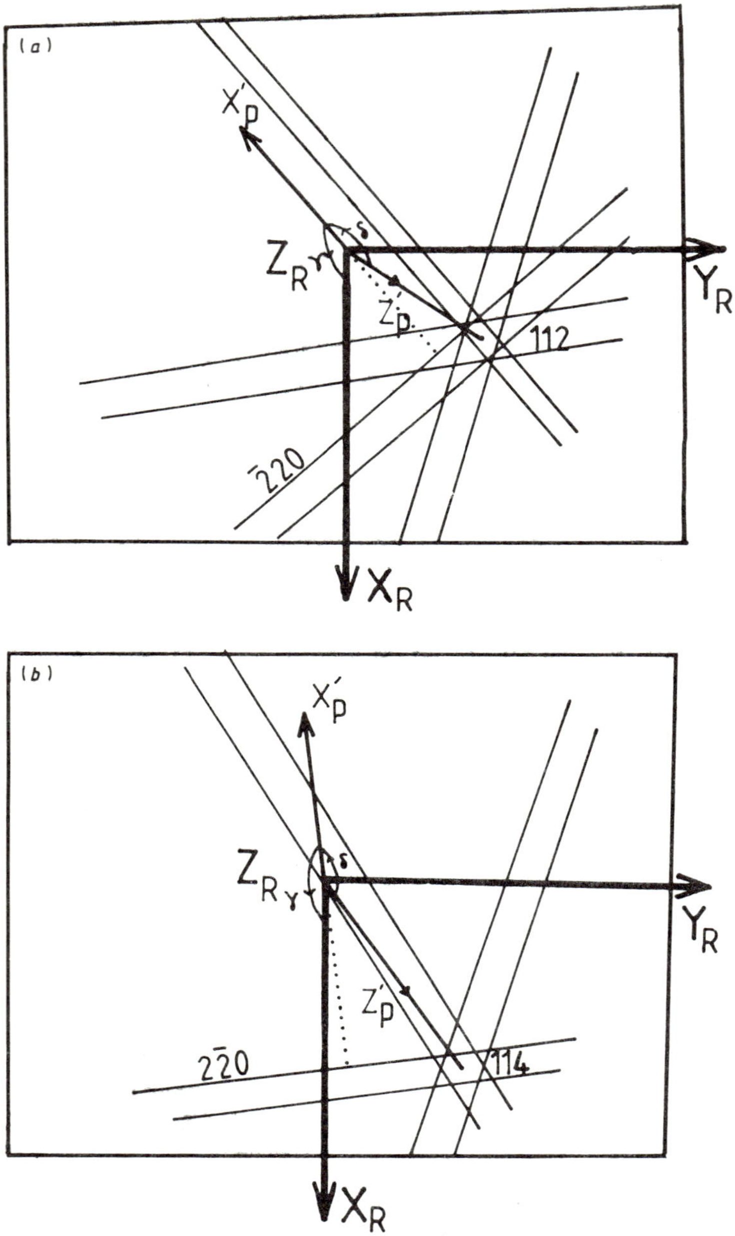

Figure 5.4 Schematic copies of the features from figure 5.3a and b which are needed to calculate an orientation for each pattern. The symbols have the same meanings as in figure 5.1

The matrix **P** for grain 1 is

$$\mathbf{P}_1 = \begin{bmatrix} -2/\sqrt{8} & 2/\sqrt{8} & 0 \\ -1/\sqrt{3} & -1/\sqrt{3} & 1/\sqrt{3} \\ 1/\sqrt{6} & 1/\sqrt{6} & 2/\sqrt{6} \end{bmatrix} \tag{5.6}$$

Having set up the pattern system in crystal coordinates, the second step is to rotate it onto the specimen-reference system (Young *et al,* 1973). This requires a knowledge of the three Euler-type angles, α,β,γ. α and β are a function of the distance Z'_P on the diffraction pattern (which is labelled on figures 5.1 and 5.4), the camera length L and the angle δ from Z'_P to X'_P which is negative if measured anticlockwise. α and β are given by

$$\alpha = \sin^{-1}[(D.\sin\delta)/(D^2 + L^2)^{1/2}] \tag{5.7a}$$

$$\beta = \tan^{-1}[(D.\cos\delta)/L] \tag{5.7b}$$

The sign of α and β is given by inspection of the sense of the rotation, with a negative sign corresponding to an anticlockwise rotation. Here α_1, β_1, α_2 and β_2 are -2.6°, -7.0°, 4.1° and -9.6° respectively. The third rotation, γ, is now about Z_R and is the angle measured from X'_P to X_R, denoting an anticlockwise rotation by a negative sign. Here γ_1 is -139° and γ_2 is -173°.

The total rotation of the pattern system to the specimen-reference system is given by the product of the following three matrices:

$$\mathbf{R}_x = \begin{bmatrix} 1 & 0 & 0 \\ 0 & \cos\alpha & -\sin\alpha \\ 0 & \sin\alpha & \cos\alpha \end{bmatrix} \tag{5.8a}$$

$$\mathbf{R}_y = \begin{bmatrix} \cos\beta & 0 & \sin\beta \\ 0 & 1 & 0 \\ -\sin\beta & 0 & \cos\beta \end{bmatrix} \tag{5.8b}$$

$$\mathbf{R_z} = \begin{bmatrix} \cos\gamma & -\sin\gamma & 0 \\ \sin\gamma & \cos\gamma & 0 \\ 0 & 0 & 1 \end{bmatrix} \tag{5.8c}$$

The multiplication must be performed in the following order:

$$\mathbf{R} = \mathbf{R_z R_y R_x} \tag{5.8d}$$

The orientation matrix, i.e. the rotation of the crystal system to the specimen-reference system, is then given by

$$\mathbf{A} = \mathbf{RP} \tag{5.9}$$

which is equivalent to equation 5.1. Finally the misorientation matrix is calculated using equation 5.5a and the 24 symmetry-related variants are derived in turn using equation 2.20 to find the solution which corresponds to the maximum trace in the matrix:

$$\mathbf{M} = \begin{bmatrix} .693 & -.641 & .377 \\ .331 & .736 & .591 \\ -.641 & -.285 & .713 \end{bmatrix} \tag{5.10}$$

We can then derive the angle/axis of misorientation from **M** using equations 2.8 and 2.9, which gives 55.9°/0.645,0.535,0.539. This is the lowest angle solution for the angle/axis of misorientation since the trace in equation 5.10 is maximised.

5.3.2. 'Triangulation' method

If a chosen pole in the specimen is the beam direction, which coincides with the direction normal to the grain surface, then the indices of it can be obtained by triangulation. The method involves setting up three simultaneous equations based on the scalar product between the vectors defining each of three poles and the vector defining ***B***. The three unknowns are then the direction cosines of $\boldsymbol{B}$. The angles Θ_1, Θ_2, Θ_3 are obtained using equation 5.2 where the camera length has already been determined. The application of the triangulation method to grain 2 (figures 5.3b and 5.4b) is illustrated in figure 5.5. The poles chosen are 114, 213 and 103. The equations to obtain the indices of B in the coordinate system of grain 2, $u_B v_B w_B$, are:

$$u_B + v_B + 4w_B/18^{1/2} = \cos\Theta_1 \qquad (5.11a)$$

$$2u_B + v_B + 3w_B/14^{1/2} = \cos\Theta_2 \qquad (5.11b)$$

$$u_B + 3w_B/10^{1/2} = \cos\Theta_3 \qquad (5.11c)$$

Since $u_B v_B w_B$ contains only two independent variables, one of the three simultaneous equations may alternatively be substituted by $u^2+v^2+w^2 = 1$. Hence only two poles need to be identified and two angles measured. However, use of three poles where possible gives a result which is inherently more accurate. Alternatively, the triangulation method can be adopted to measure the angle from $\boldsymbol{B}$ to three non-coplanar Kikuchi bands which can be indexed by reference to a stored look-up table (Heilmann *et al,* 1982; Schwarzer, 1990).

Equation 5.11 yields a value for $u_B v_B w_B$ in grain 2 of 0.414, 0.173, 0.894. A convenient Kikuchi line is then chosen in the pattern; if this line is *hkl* then the matrix which describes the rotation of the crystal system to the pattern system is formulated thus (Heilmann *et al,* 1982):

$$\mathbf{P} = \begin{bmatrix} (\boldsymbol{B} \times hkl) \times \boldsymbol{B} \\ \boldsymbol{B} \times hkl \\ \boldsymbol{B} \end{bmatrix} \qquad (5.12)$$

In grain 1 and 2 hkl is chosen to be $2\bar{2}0$ and $\bar{2}20$ respectively. The only further angle required to rotate the pattern system onto the specimen-reference system is γ, which is measured as shown in figure 5.4. Hence for this example γ_1 is 40° and γ_2 is 0°. Thus the matrix **R** which transforms between the crystal system and the specimen-reference system consists only of equation 5.8c and the final orientation matrix is given by equation 5.9. Using the triangulation method the misorientation between grains 1 and 2 is 55.2°/0.620, 0.566, 0.533, which gives good agreement with the value obtained by the 'pole and line' method described in section 5.3.1. The 'triangulation' method, rather than the method involving Euler angle rotations which was described in the previous subsection, forms the basis for the EBSD orientation measurement software routine.

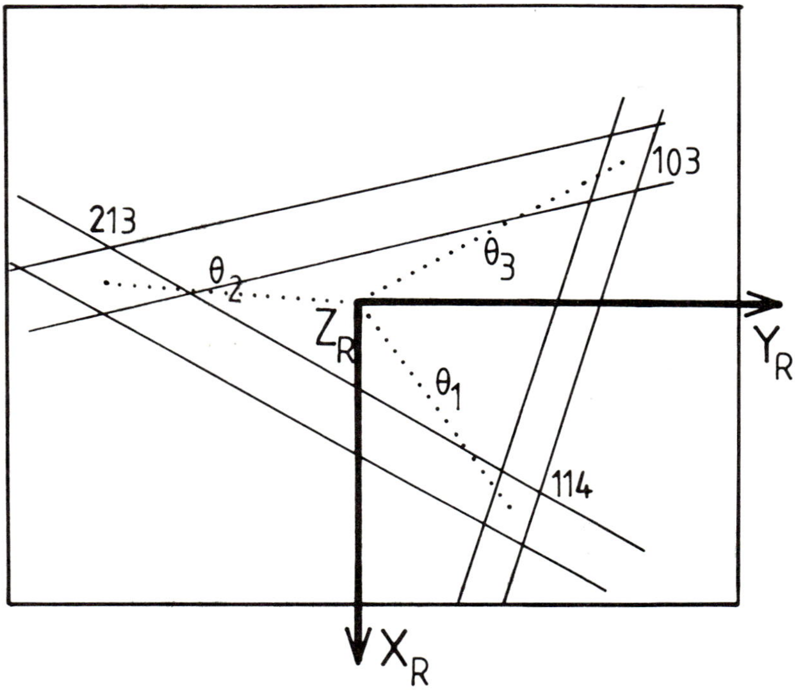

Figure 5.5 Illustration of the 'triangulation' method for calculating an orientation with reference to the diffraction pattern in figure 5.3b (grain 2).

5.3.3 Stereographic method

The angle/axis of misorientation can be obtained by stereographic construction, which was described briefly in section 2.2.2. The method requires the knowledge of three pairs of directions (non co-planar) which have the same indices in both grains (Randle and Ralph, 1986). The most convenient choice is the crystal axes. These are plotted on a pair of stereograms, one for each grain, by transferring the beam direction to the standard stereogram which has axes 100, 010 and 100 and then rotating ***B*** into the centre. The two stereograms centered on ***B*** defined in grains 1 and 2 respectively are then combined and rotated about ***B*** (which is now common to both grains) by the angle γ defined on figure 5.1 and in section 5.3.1 with the result that the $\boldsymbol{X_R}$ and $\boldsymbol{Y_R}$ axes from each grain are now aligned parallel. The crystal axis pairs and reference axes for diffraction patterns from grains 1 and

2 (figure 5.3) which are being used as an example in this section are shown on the stereogram in figure 5.6 (see also figure 5.8).

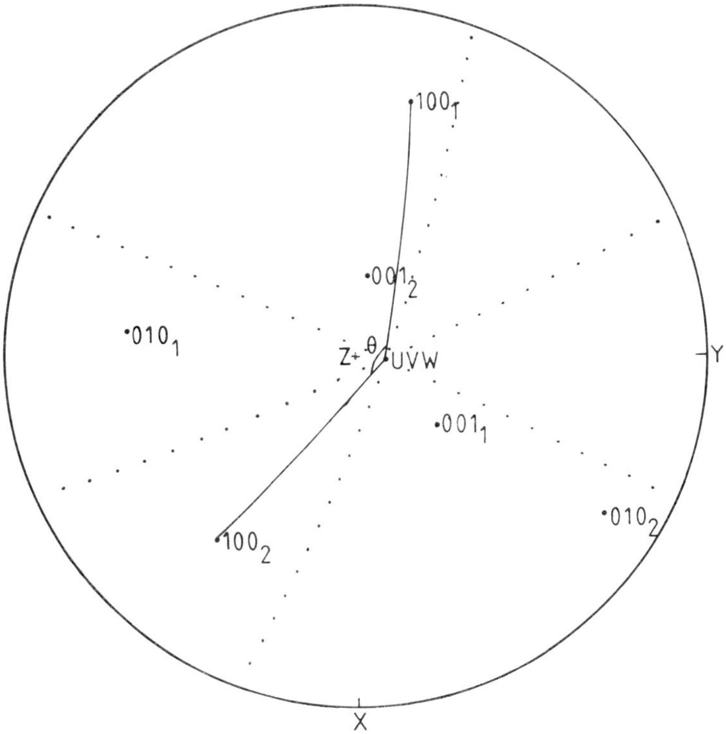

Figure 5.6 Use of the stereogram to obtain a misorientation, using the diffraction pattern pair in figure 5.3 as an example.

Because the axis of misorientation has the same Miller indices in both grains, it lies on a zone (great circle) which is equidistant from each pair of crystal axes. These three zones can be constructed on the stereogram and their intersection gives *UVW* as shown on figure 5.6. To define *UVW* by the intersection of three zones rather than two is a check against plotting errors and increases the accuracy of the determination. The angle of misorientation is also shown on figure 5.6 and is best obtained by measuring the angle between poles of great circles which contain (*UVW*, 100_1) and (*UVW*, 100_2) respectively. The procedure is repeated for 010 and 001 pairs so that Θ is averaged from the three measurements. Information about stereographic manipulations can be found elsewhere (Johari and Thomas, 1970; McKie and McKie, 1974).

The data plotted in figure 5.6 give a value of 150°/0.331 0.326 0.895 for Θ/UVW. This solution for Θ/UVW, obtained directly off the stereogram, corresponds to the misorientation matrix obtained by the analytical method which was described in the previous subsections. To obtain the lowest angle solution the angle/axis must be re-expressed in matrix form using equation 2.11, multiplied by the each of the 24 symmetry relations as in equation 2.20 and reformulated as an angle/axis using equations 2.8 and 2.9 for the matrix which has the maximum trace. This procedure gives Θ_{min}/UVW as 58.3°/0.497 0.606 0.621 which is in good agreement with the value obtained by the previous two analytical methods, bearing in mind the accuracy limitations associated with plotting and measurement on the stereogram.

The stereographic method is unlikely to be of use for the processing of even moderate quantities of diffraction data, since a computer program will perform the analysis more quickly and accurately. However, the stereogram is a graphic means of demonstrating the misorientation concept and also it provides a means of checking the analytical method.

5.4 CATEGORISATION OF GRAIN BOUNDARY GEOMETRY

Having obtained a misorientation from two diffraction patterns, the next step is usually to categorise it as an LA, CSL or CAD type. The angle/axis format is the description of the misorientation which lends itself best to physical interpretation. For example when the misorientation is expressed in angle/axis form, LAs (or any other angle category, e.g. figure 6.2b) are immediately apparent. The most commonly accepted cut-off for LA categorisation is 15° (Dechamps *et al,* 1987). Misorientations can also be expressed as Euler angles. In general this is a more impractical approach for GBs than that of Θ/UVW, and tends to be used only in connection with misorientation distribution functions, MODFs (see section 6.4).

From a knowledge of Θ/UVW for both the experimental and the CSL case (Table 3.3 and 3.4) it is possible to identify GBs which are close to CSLs. For the example in the previous subsection it is clear that Θ/UVW, which was 55.9°/0.645, 0.535, 0.539, is close to the value of Θ/UVW for a $\Sigma = 3$ CSL, 60°/111. However, we need to specify this closeness quantitatively which can be done by several analytical methods.

One method compares analytically the misorientation matrices for the experimental and CSL cases, $\mathbf{M}_{EXP}$ and $\mathbf{M}_{CSL}$ respectively (Dechamps *et al,* 1987; Randle and Ralph, 1988c). The difference matrix $\mathbf{M}_D$ is given by

$$\mathbf{M}_D = \mathbf{M}_{EXP}\,\mathbf{M}_{CSL}^{-1} \qquad (5.13)$$

The angular deviation v between $\mathbf{M}_{EXP}$ and $\mathbf{M}_{CSL}$ is given by the trace of $\mathbf{M}_D$ according to equation 2.8. $\mathbf{M}_{EXP}$ and $\mathbf{M}_{CSL}$ refer to the same symmetry related solution, usually the one containing the lowest angle.

The angular deviation from exact CSL matching can also be obtained by calculating the average of the three angles between the 100, 010 and 100 directions in both $\mathbf{M}_{EXP}$ and $\mathbf{M}_{CSL}$, i.e. the angles between the first, second and third columns in these two matrices (Randle and Ralph, 1986). Another method for obtaining v is based on geometrical considerations (Sharko, 1983). Here v is given by

$$v = (\mid \Theta_{EXP} - \Theta_{CSL} \mid ^2 + (2\Theta_0 \sin(\Theta_{EXP}/2))^2)^{1/2} \quad (5.14)$$

where Θ_0 is the angular difference between $(UVW)_{EXP}$ and $(UVW)_{CSL}$.

All of the three methods for obtaining v give the same result, which for the example in the previous section is 4.5°. The specification of a maximum allowable deviation that an experimental GB can have and still be classed as a CSL, v_m, was discussed in section 3.2.3. Almost always the 'Brandon criterion' is adopted to give v_m. Using the Brandon criterion in the present example of a $\Sigma = 3$ GB gives the relative deviation

$$v/v_m = 4.5°/8.66° = 0.52 \quad (5.15)$$

The deviation from a CAD GB can be obtained in the same way as for a CSL, substituting the CAD misorientation axis for the experimental one (Warrington and Boon, 1975; Randle and Ralph, 1988c). For example, the nearest CAD axis for a GB with Θ/UVW 33.2°/ 0.982 0.159 0.105 is 100. Equation 5.13, 5.14 or the 'columns' method can then be used to find the deviation between the experimental Θ/UVW and 33.2°/100, which is 6.3°. The upper limit for a CAD GB with $\pi = 4$ (200 planes, see section 3.3.1 and Table 3.7) is 10.1° for fcc material, and so this GB can be classified as a CAD.

The most efficient way to check for a CSL is to store the data for all CSL misorientations up to a predetermined Σ limit in a look-up table as part of the computer program for determining the misorientation. The lowest angle misorientation for an experimental GB is calculated and compared with all the CSL misorientations, which are also stored as the lowest angle form. A similar procedure can be applied for CADs. Table 5.1 summarises all the data processing steps from the 'raw' diffraction pattern to the categorisation of the GB geometry.

5.5 CALCULATION OF THE NORMAL TO THE GRAIN BOUNDARY PLANE

The input data needed to calculate the GB plane normal are:

1. The orientation of each neighbouring grain with respect to the specimen-reference axes $X_R Y_R Z_R$;
2. The angle α that the trace T of the GB makes with the X_R axis on the specimen surface;
3. The angle of inclination φ that the GB plane makes with the Z_R direction.

The specification of $X_R Y_R Z_R$, T, α and φ is shown on figure 5.7. The measurement of these parameters was discussed in sections 4.3.2 and 4.4.3 for methods using both the TEM and SEM. The same procedure for calculating the GB plane normal can be used for TEM and SEM once the relevant parameters have been obtained. It is important to ensure that the GB inclination angle, φ, is defined as that shown on figure 5.7, i.e. measured between the Z_R direction and the GB plane. The TEM foil tilting method refers to the complement of φ, which is φ', and the SEM 'two-surface' method gives the inclination measured from the $-X_R$ direction, ß as shown in figure 4.12 (Randle and Dingley, 1989, 1990a). The GB inclination is referred to the X_R axis in this method because the two adjoining surfaces are viewed in the microscope along their common X_R axis, and so it is convenient to choose the X_R axis as a datum line for angular measurements. The relationship between φ and ß for both $ß < 90°$ and $ß > 90°$ is shown on figures 5.7a and b respectively. When $ß = 90°$ $\varphi = 0°$.

It is also important to choose the sense of φ correctly. If the rotation about T between the GB plane and the Z_R axis is anticlockwise, φ is defined as negative, which is the case illustrated in figure 5.7a. Conversely a clockwise rotation about T gives a positive φ (figure 5.7b). α is measured from the X_R axis to the trace direction T, and so is always positive.

Three sets of simultaneous equations are needed to solve for the GB plane normal indexed in both grains (Randle and Dingley 1992). The first set gives the direction cosines of T:

$$T.X = \cos\alpha$$

$$T.Y = \sin\alpha \qquad (5.15a)$$

$$T.Z = 0$$

where X, Y, Z are the crystal directions with reference to the specimen-reference axes $X_R Y_R Z_R$ firstly for grain 1 and secondly for grain 2. Next the

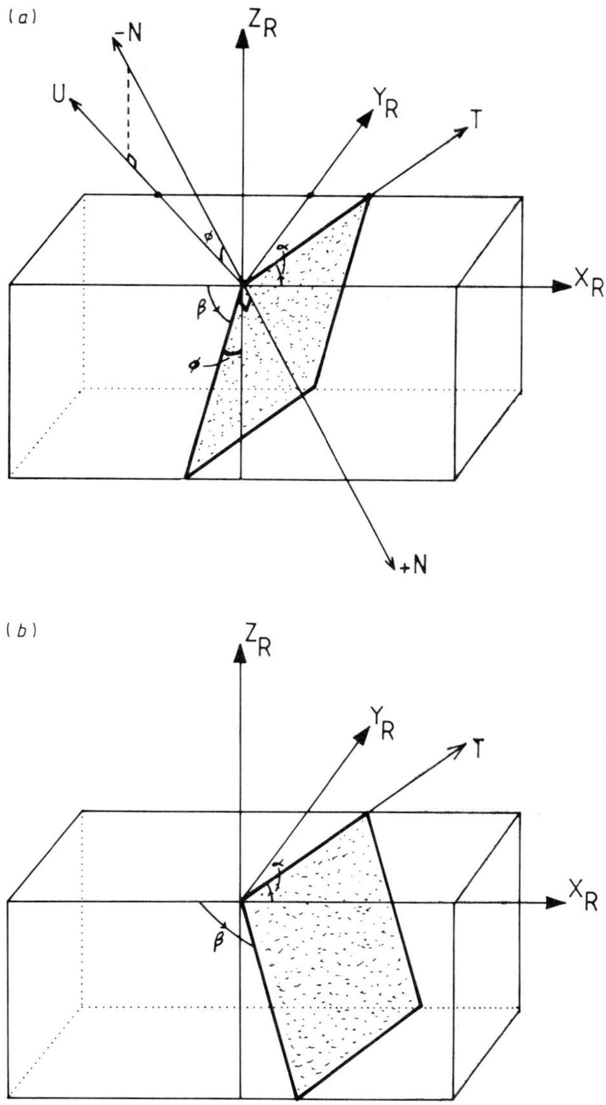

Figure 5.7 Specification of the parameters needed to measure the GB plane normal, ***N***. ***U*** is a direction perpendicular to the GB trace direction ***T***, with ***T*** and ***U*** co-planar. φ is measured in either a serial sectioning or a thin specimen method, and α, ß are measured in the 'two-surface' method. In (a) φ is defined as negative, and in (b) it is positive.

direction cosines of U, a direction which is on the specimen surface (i.e. in the $X_R Y_R$ plane) and perpendicular to T, are found:

$$U.Y = \cos \alpha$$

$$U.T = 0 \qquad (5.15b)$$

$$U.Z = 0$$

X_R, T, Y_R and U are co-planar as shown on figure 5.7a. Since the angle between U and N is the angle of inclination φ we can now obtain the direction cosines of N:

$$N.U = \cos \varphi$$

$$N.Z = \sin \varphi \qquad (5.15c)$$

$$N.T = 0$$

We will consider three examples of determinations of N. The first was determined by EBSD, and refers to the labelled GB in figure 4.11, which shows a pure iron specimen. Here α is 141° and φ is 25°, as determined by removing 54μm of the surface which was measured using a hardness indent for calibration as shown in figure 4.11. The GB is a $\Sigma = 3$ CSL, with a relative deviation $(v/v_m) = 0.33$. Using the method described above, we obtain the following for N indexed in both grains:

$$N_1 = .423 \ -.423 \ -.799 \qquad (5.16a)$$
$$N_2 = -.719 \ .276 \ -.636 \qquad (5.16b)$$

N_1 is 1.5° from <211> and N_2 is 4.8° from <221>. According to Table 3.6 this does not correspond to an asymmetrical tilt in the $\Sigma = 3$ system. The plane normals are, however, close to quite low index planes in the lattice. Frequently in polycrystals the GB aligns parallel to low index lattice planes, particularly in one grain only (Randle and Ralph, 1988f; Randle, 1989, 1991c; Carter, 1988; Wolf, 1988; Merkle and Wolf, 1992).

Figure 5.8 shows the determination of N on a stereogram. T is measured from X_R along the primitive, which represents the plane defined by $X_R Y_R$. U

is measured 90° from $\boldsymbol{T}$ on the primitive, and then to locate N we rotate $\varphi°$ from $\boldsymbol{U}$ along a great circle of which $\boldsymbol{T}$ is the pole. Care must be taken to perform the rotation in the correct sense. The crystal axes of the two interfacing grains relative to the specimen axes are obtained from the rows of the orientation matrix (following the standard convention in equation 5.4) and plotted on the stereogram. The inverse cosine of each direction cosine which represents a crystal axis gives the angle of the crystal axis from $\boldsymbol{X_R}$, $\boldsymbol{Y_R}$ and $\boldsymbol{Z_R}$. The direction cosines of N can then be obtained relative to grain 1 and grain 2 from the cosines of the angles between N and 100, 010 and 100 respectively.

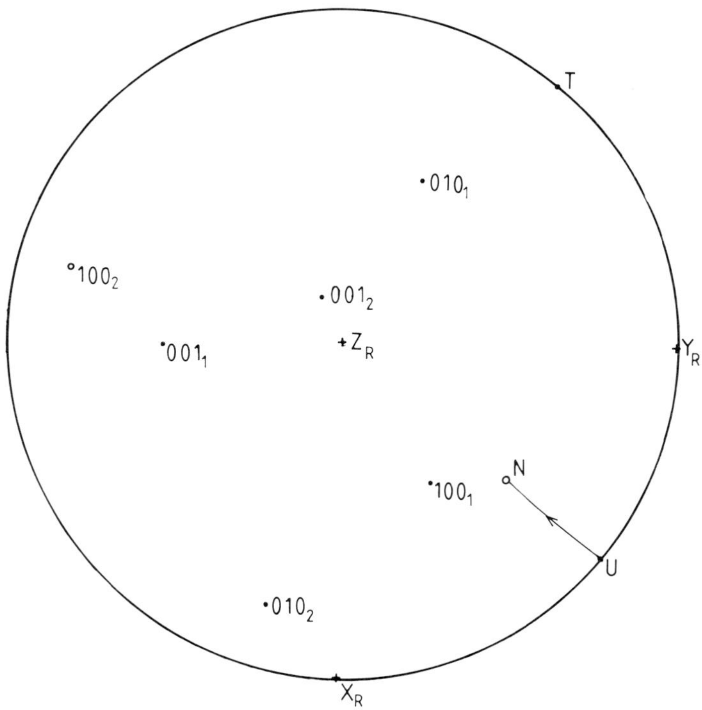

Figure 5.8 Stereographic interpretation of the calculation of a GB plane normal using data from one of the GBs shown in figure 4.11. The symbols have the same meaning as in figure 5.7.

The second example where all five degrees of freedom for a GB have been determined was obtained in the TEM from an austenitic steel (Randle, 1989). Figure 5.9 shows a dark field image of a $\Sigma = 33c$ and a $\Sigma = 3$ GB, $v/v_m = 0.61$ and 0.72 respectively, and grains labelled 1,2,3. There are precipitates on the $\Sigma = 3$ GB. (Note that in the dark field imaging mode it is difficult to recognise the top and bottom of the GBs.) The GB normals are:

$\Sigma = 33c$:

$$N_1 = .616\ \text{-}.500\ \ .588 \tag{5.17a}$$

(5.0° from <111>, 4.5° from <554>)

$$N_2 = .669\ \ .484\ \ .559 \tag{5.17b}$$

(7.6° from <111>, 1.5° from <554>)

$\Sigma = 3$:

$$N_2 = \text{-}.743\ \ .669\ \ 0 \tag{5.17c}$$

(3.0° from <110>)

$$N_3 = .407\ \text{-}.515\ \ .754 \tag{5.17d}$$

It is estimated that a deviation from the exact GB plane indices of 5-7° can be accommodated at the GB (Wolf, 1985) and the near <554> planes in the $\Sigma = 33c$ GB fall into this category. These planes are twin planes in the $\Sigma = 33c$ system (section 3.2.4) and so the GB is a symmetrical tilt GB (section 3.2.5) which is described in the interface plane scheme as $[5\bar{5}4][554]\ 0°$. Furthermore, N_1 and N_2 are both quite near to <111>, which is the closest packed plane in the lattice. The $\Sigma = 3$ GB is classed as general in the interface-plane scheme because it is not a STGB, ATGB or TWGB. One of the interfacing planes is close to 220, a low-index plane in the lattice.

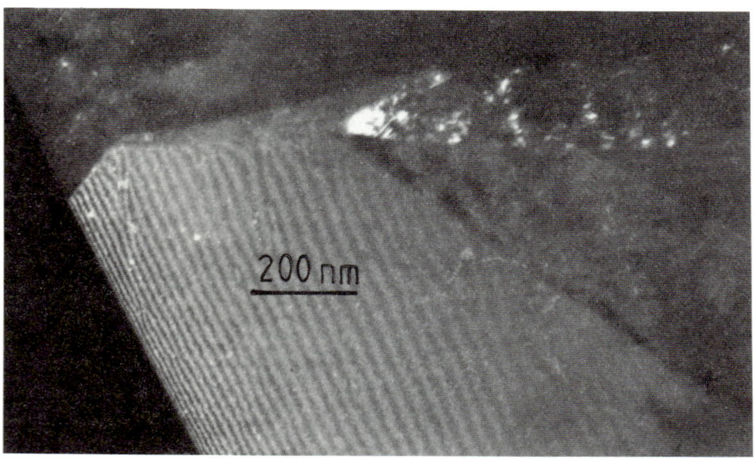

Figure 5.9 Dark-field image of a $\Sigma = 3$ and a $\Sigma = 33c$ GB in an austenitic steel. There are precipitates in the $\Sigma = 3$ GB. The GB plane normals of both these GBs have been measured.

Finally we show an illustration of a case where the experimental measurements for N were made using the 'two-surface' method. The specimen is pure nickel, the GB is a $\Sigma = 3$, $v/v_m = 0.17$. α and β are 106° and 146° respectively. Thus the GB is inclined as shown on figure 5.7b. The GB normals are:

$$N_1 = -.595\ \ .547\ \ .589 \quad (2.1° \text{ from } <111>) \tag{5.18a}$$

$$N_2 = .952\ \ .254\ \ .172 \quad (3.8° \text{ from } <511>) \tag{5.18b}$$

Thus this GB is an ATGB which in the interface plane scheme is described as $[\bar{1}11][511]\ 0°$.

6

DATA REPRESENTATION AND DISPLAY

6.1 INTRODUCTION

The most appropriate mathematical form for calculations concerning GB geometry is the misorientation matrix, **M**, and a vector which describes the direction of the GB plane normal, N. The matrix formulation, which is nine numbers, overdefines the misorientation, since only three independent variables are actually needed. However, the matrix form is particularly convenient for the computation involved in data processing, as described in the previous chapter. As it stands, a misorientation matrix does not lend itself readily to representation of data in some physically meaningful way. This is in contrast to an *orientation* matrix where important directions in the specimen, typically the rolling direction and rolling plane, are the first and last columns of the rotation matrix respectively (using the standard convention, see section 5.2) usually converted to the nearest Miller indices (Bunge, 1987,1988). The 'ideal orientation' representation of **M** is clearly not physically meaningful for GBs. It is necessary to convert **M** into other forms so that the data can be interpreted. To recapitulate from chapter 2, the most useful embodiments of **M** are

1. The angle/axis of misorientation (see section 2.2.2)
2. The Euler angles (see section 2.3.2).

If N has been measured in addition to **M** the most common options for representation are

3. The 'misorientation scheme', Θ/UVW, N (see section 2.2.2)
4. The 'interface-plane scheme', N_1, N_2, φ (see section 2.2.1)

The orientation of each neighbouring grain relative to the specimen axes may also be included as input parameters, so that, for example, **M** and/or N could alternatively be expressed relative to the orientation of each grain and to each of the specimen axes. A further consideration is that **M** can be expressed in 24 different symmetry-related ways (see section 2.4) which in turn gives 24 equivalent solutions of (1) to (4) above.

The representation of a population of GB parameters is therefore the re-expression of **M** and *N* in a manner which is appropriate for classification and thus interpretation of the data. Having chosen a suitable method of representation, we also need to display it in a way which facilitates visual appraisal of the salient features of the data. In practice the terms 'data (re)presentation' and 'data display' are used interchangeably. There are essentially two types of display: one embodies only the statistical distribution of the misorientation and/or plane normal data, and the other includes how individual GBs are distributed in space. Some of the methods used to display GB statistical data are analogous to those employed for orientations, e.g. representation in Euler space. As far as spatial distributions are concerned, typically a 'mapping' technique is used whereby the GB parameters are superimposed on a micrograph, or diagram taken from a micrograph, which shows the connectivity of the GBs (section 6.5).

If the misorientation only is measured, the task of data display reduces to that of presentation of three independent parameters which is usually accomplished by presentation of the data in a series of two-dimensional sections. Where additional parameters have been measured, the data are still usually displayed in sets of two parameters at a time for the sake of convenience on the printed page. For example, the axis of misorientation (two independent parameters) could be displayed relative to *one* of the specimen axes by using a sectionalised display.

In this chapter we will examine in detail each type of representation and display, and also how departures from random distributions are recognised and quantified. The representation methods can be categorised according to the space in which they reside. The three most important spaces, which are all based on orthogonal axes, are:

1. Stereogram-based space (section 6.2), where typically the misorientation axis is displayed stereographically in the plane of the page and the misorientation angle is displayed on an axis perpendicular to it. Frequently the misorientation axis but not the misorientation angle is displayed whereupon only a two-dimensional presentation, i.e. the stereogram, is needed. Stereogram-based space is also used to display GB plane normals.
2. Rodrigues-Frank (RF) space (section 6.3), where the Rodrigues vector representations of misorientations are plotted. A Rodrigues vector has three components and is a function of Θ and *UVW*.
3. Euler space (section 6.4), which is used to plot a misorientation when it is represented as Euler angles. It is common to use this space not only to display experimentally measured individual misorientations, but also to display misorientations derived on a probability basis from the distribution of

orientations. This is called the misorientation distribution function (MODF or MDF).

Data display of any of these three-dimensional spaces is by one of the following:

1. A series of sections parallel to one of the axes;
2. Projection of all the data points onto one plane;
3. A perspective drawing of the three-dimensional space;
4. Stereo pairs;
5. Video display;
6. Holographic displays.

The first two in this list, sections and projections, are the most widely used. A further choice to be made is whether to display data as discrete points or to smooth it into contours. Software is available which transforms from a distribution of individual data points to density contours. Each misorientation may be 'weighted' according to the GB projected lengths in a measurement section and/or the metric of the space (see section 6.4). Usually, small data populations are more conveniently represented as discrete points and conversely the features of large data sets, or when the distribution of points is particularly dense such that detail is obscured, can be best appraised as contours. There are examples of both in this chapter.

In addition to these specific categories for statistical data presentation, more general means such as frequency histograms, charts, graphs etc. are very often used to display and highlight aspects of the data, especially in terms of 'special' GB classification.

6.2 STEREOGRAM-BASED SPACE AND 'SPECIAL' GRAIN BOUNDARY PROBABILITIES

The most common plot of this type is the 'misorientation axis distribution,' or 'grain misorientation texture' figure (Randle and Ralph, 1987b, 1988a,f,g; Randle and Brown, 1989) where *UVW* for the lowest angle solution of the 24 possible selections, the 'disorientation', is standardised so as to lie in a single unit triangle of the stereogram. Frequently the misorientation angle is not represented so that only a single figure is needed, in analogy to an inverse pole figure. The stereogram is based on the crystal axes and thus the figure has the same form as an inverse pole figure. Figure 6.1 shows an example from an austenitic steel specimen. When the number of data points is small, each can be identified individually and labelled as appropriate, e.g. with a Σ-value, as in figure 6.1 and/or correlated with the spatial distribution.

Figure 6.2a shows a type of misorientation axis distribution where the misorientation axes are plotted relative to the specimen axes rather than those of the crystal. This representation of the GB misorientation data is thus analogous to a pole figure rather than an inverse pole figure, and is less

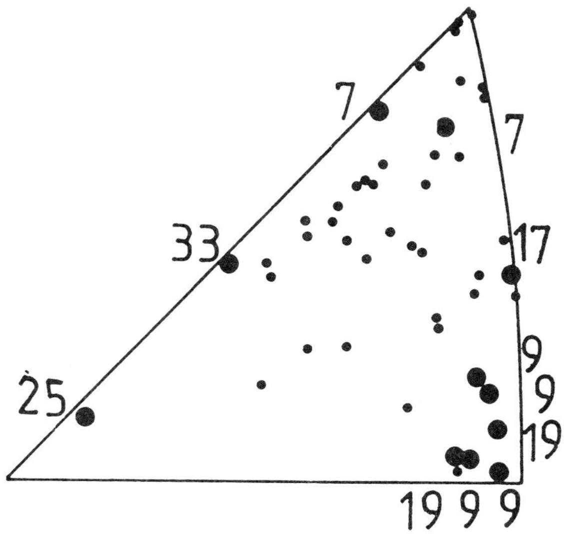

Figure 6.1 Disorientation axes (i.e. axes of misorientation for the lowest angle solution) represented in a unit triangle of the stereogram with CSLs as larger symbols labelled with a Σ-value. The specimen is a nickel-base alloy (Randle et al, 1988).

common than the type of distribution shown in figure 6.1 because usually misorientations are measured without reference to the external axes, such as an axis or axes of applied stress (Palumbo et al, 1991). The data in figure 6.2 are from 95% reversed rolled copper. The misorientation axes show a strong preference for the specimen normal direction, and clearly this choice of display method highlights this particular misorientation texture. The distribution of misorientation angles is shown separately on figure 6.2b. The random distribution is also included on this figure as a dashed line. The type of plot shown in figure 6.2b is commonly used where the distribution of misorientation angles is of more interest than that of misorientation axes, e.g. where there is a large proportion of low angle GBs.

A sectioned three-dimensional misorientation axis distribution is shown in figure 6.3 where Θ is represented on a third axis, perpendicular to the stereogram, in 10° sections. This third axis extends to the largest possible disorientation angle, 62.8°. Figure 6.3 will be referred to again in section 6.3.

The total volume of the three-dimensional misorientation space is cylindrical, with the misorientation angle represented along the axis of the cylinder. Only 1/24th of the cylinder is used in figure 6.3 because the misorientation axes are standardisd to a single unit triangle. Reference to figure 6.4 shows the form of cylindrical misorientation space for Θ/*UVW* expressed as polar coordinates where the complete stereographic projection is used. The metric (i.e. the 'volume-trueness') of misorientation space is not constant, but varies according to the density of poles D as a function of their spherical coordinates p_1, p_2, p_3 (Warrington, 1980):

$$D(p_3, p_2, p_1)\mathrm{d}p_3\mathrm{d}p_2\mathrm{d}p_1 = 1/(2\pi^2)\sin^2(p_3/2)\sin p_2\, \mathrm{d}p_3\mathrm{d}p_1 \qquad (6.1)$$

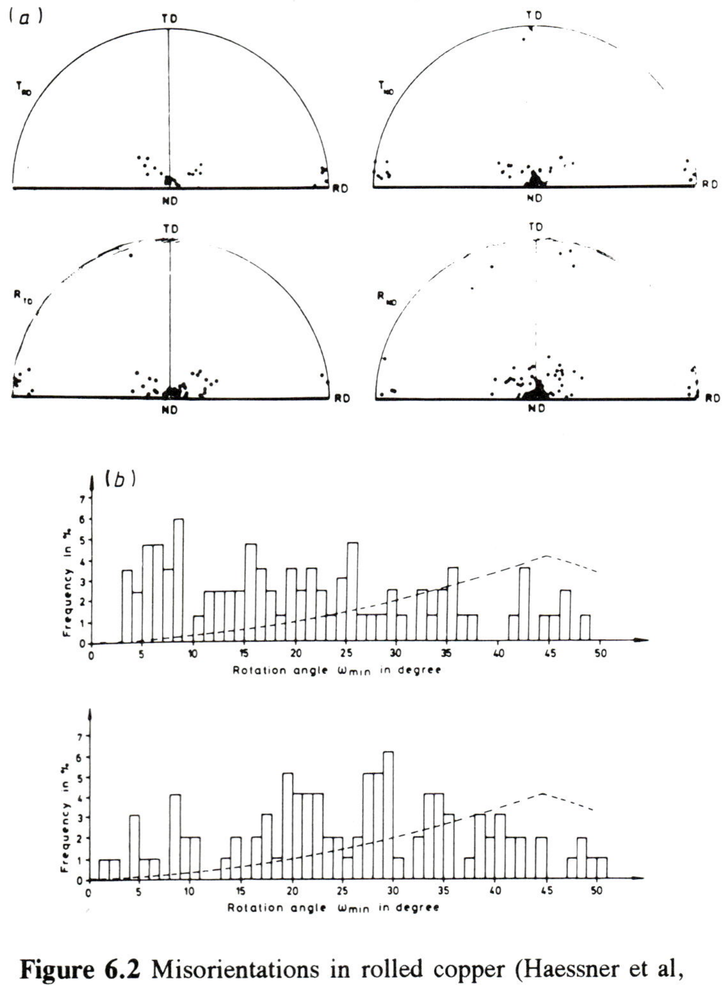

Figure 6.2 Misorientations in rolled copper (Haessner et al, 1983. (a) Distribution of disorientation axes relative to specimen axes. (b) Accompanying frequency distribution of minimum misorientation angles. The distribution of angles for the case of statistically random misorientations is shown as a broken curve.

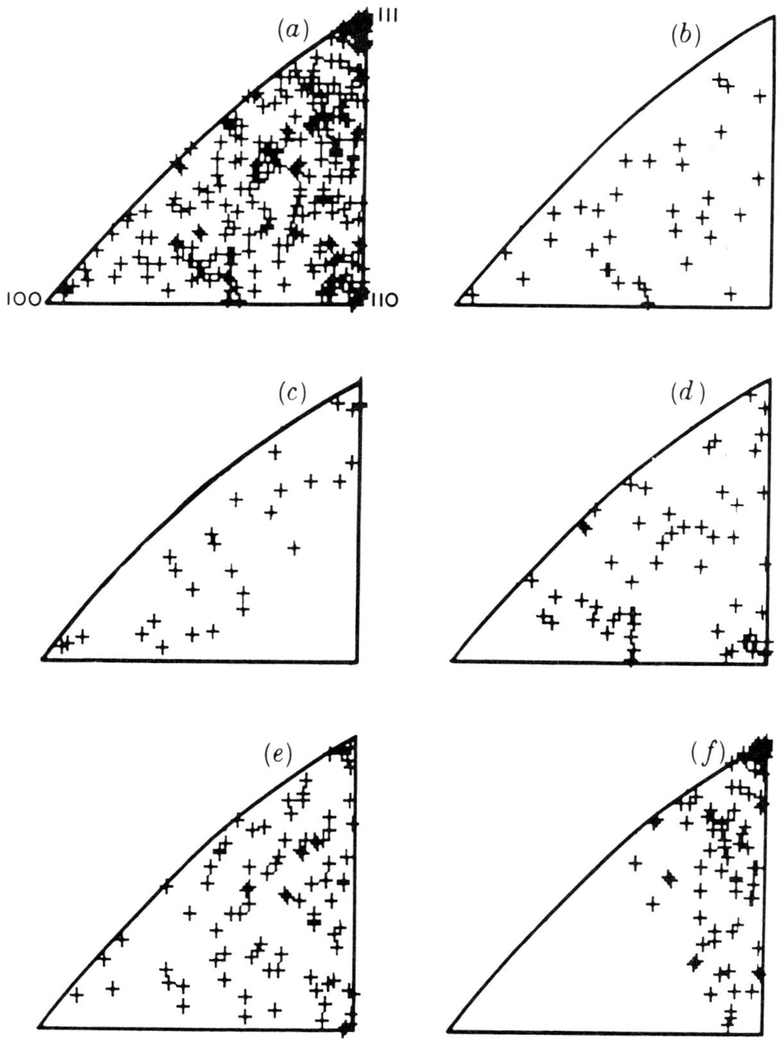

Figure 6.3 Disorientation axes from an fcc steel sample represented in a stereographic triangle (a) for the whole data set irrespective of angle and (b-f) for specific ranges of misorientation angle: (b) <20°, (c) 20-30°, (d) 30-40°, (e) 40-50°, (f) >50°. Thus (b-f) 'stack' to form a three-dimensional figure (Randle, 1990a).

Figure 6.4 shows that the deformation of the space is greatest for small misorientation angles. The use of this form of stereogram-based space is discussed further and compared to Euler space in section 6.4.

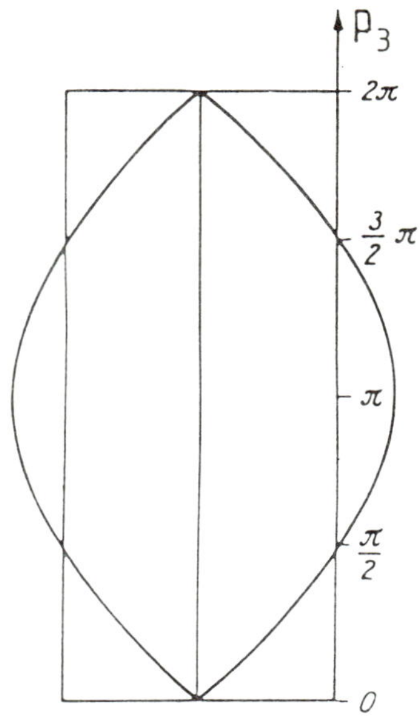

Figure 6.4 Longitudinal section through the cylindrical stereogram-based space (i.e. parallel to the misorientation angle, p_3) to show the deformation of the space (Pospiech et al, 1986).

An alternative form of stereogram-based space is a unit reference sphere where the normalised misorientation axis and angle are defined by the direction and magnitude of a vector respectively (Warrington, 1980). All misorientations project stereographically onto the equatorial plane and so the cylindrial and spherical forms of the misorientation space are equivalent with respect to misorientation axes.

Since only one Θ/UVW out out 24 possible is selected for cubic crystals, the distribution of vectors for a totally random misorientation distribution is not distributed randomly on the reference sphere. To emphasis this point, a random distribution of misorientations would give rise to a uniform distribution if the crystals were triclinic because here each individual misorientation can be described by only one Θ/UVW. The anisotropic density distribution of misorientations for the cubic case makes it difficult to assess visually when a distribution of misorientations is random. However, the probability density for UVW can be found by integrating equation 6.2 over the limits of the variables and dividing by dS, which is an element of area on the surface of the unit sphere. This probability density can then be expressed as the proportion of misorientation axes expected to lie within defined areas of the unit stereographic triangle for the lowest angle solution if the distribution is random (Mackenzie, 1964). Figure 6.5a shows these proportions. Also, the relative proportions of misorientation axes expected to fall within a small fixed angle (e.g. 5°) of 100, 111 and 110 are in the ratio 1:3.1:4.8 for a random distribution.

It is frequently of interest to assess the statistics of a GB population (Werner and Prantl, 1988), particularly in terms of the probabilities of the occurrence of geometrically special GBs. The expected CSL, LA and CAD proportions have been calculated as a corollary to the 'Mackenzie triangle' formalism shown in figure 6.5 and so it is appropriate to consider them in the present section.

For the case of grains which have randomly distributed orientations, i.e. a random texture, misorientation distributions have been generated by using a random number generator to compute pairs of neighbouring orientations. From these data the probabilities of CSLs, LAs and CADs have been collated, and the expected proportions of them are 9% (up to $\Sigma = 25$), 2% and 56% (up to $\pi = 8$) respectively (Warrington and Boon, 1975). More recently this type of simulation has been repeated including connectivity in the input orientations, that is, taking into account all the neighbours of a grain in a polycrystal (Garbacz and Grabski, 1989). This approach gives a more realistic simulation of a microstructure. The CSL fraction up to $\Sigma = 25$ remained the same at 9%, yet the distribution of CSLs within this range is different as shown in figure 6.5b. The 'connectivity' simulations give a good agreement with the expected proportions in the unit stereographic triangle as shown in figure 6.5a. As might be expected, if a preferred orientation (texture) is introduced into the simulations, the CSL proportions change greatly (Randle and Ralph, 1988d).

Figure 6.6a shows an experimentally obtained misorientation axis distribution from low carbon steel sheet. For this example the distribution is essentially random. A second example from a nickel-base alloy is shown in figure 6.6b. Here there is a greater departure from randomness, and this is assessed by expressing the proportion of poles in each region marked on figure 6.5 as 'times random'. The significance of the numbers thus obtained can be evaluated by a statistical treatment such as the Student t-test (Randle and Ralph, 1988a) or the ψ^2-test (Prantl et al, 1988).

The unit stereographic triangle is also used to show the direction of GB plane normals. Figure 6.7a shows an example from NiO where N is plotted for each grain, i.e. two normals are plotted per GB. Those planes which are from CSL GBs are marked with the Σ-value. In figure 6.7b and c, N is plotted relative to the specimen axes. The significant feature for the data in figures 6.7b and c is that the GB planes tend to align parallel to the specimen Z-axis for non-CSL GBs, and not parallel to the Z-axis for CSLs.

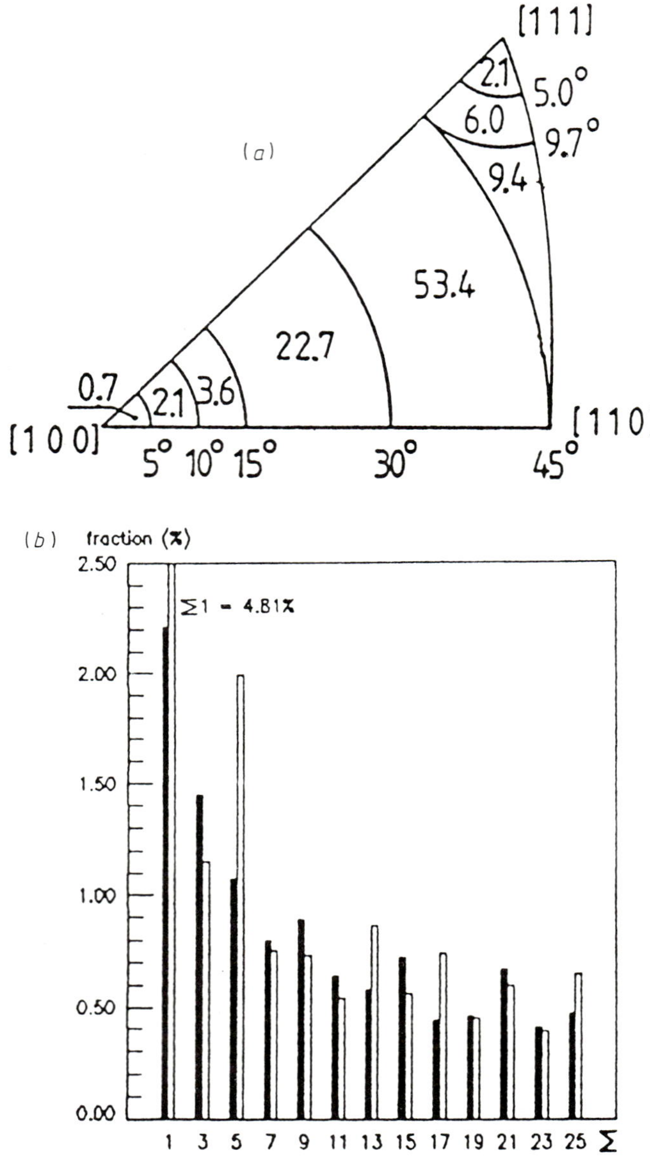

Figure 6.5 Distribution of misorientations for the statistically random case. (a) Distribution of disorientation axes in the unit triangle showing the expected percentages in each case (Mackenzie, 1964). (b) Frequencies of CSLs for both randomly selected grain pairs (open histogram) and the polycrystal case (filled histogram) (Garbacz and Grabski, 1989).

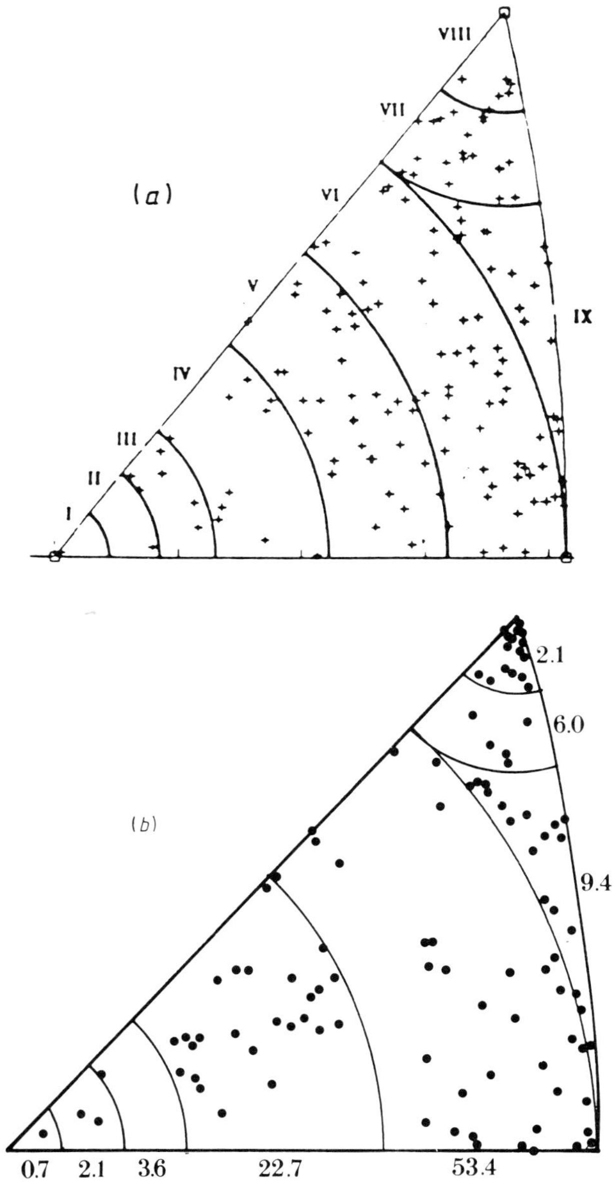

Figure 6.6 Experimental examples of data plotted in the 'Mackenzie triangle' shown in figure 6.5a. (a) Data from a bcc low carbon steel, showing an almost random distribution (Rabet et al, 1992). (b) Data from an fcc nickel-base alloy showing a greater departure from randomness than in (a), particularly with respect to axes close to 111 (Randle and Ralph, 1988a).

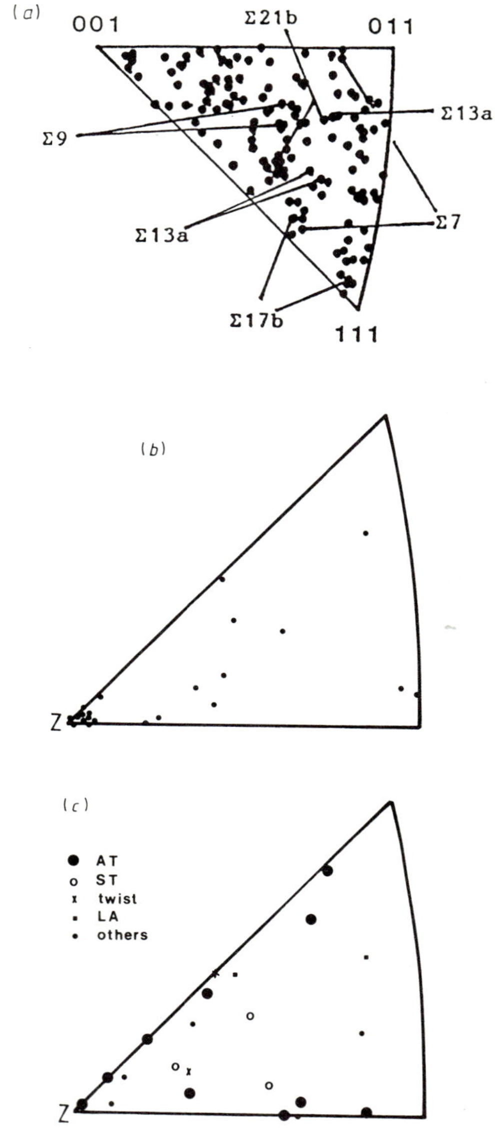

Figure 6.7 GB plane normals represented in a stereographic unit triangle. (a) data from NiO, with CSLs labelled with a Σ-value. There are two data points per GB to represent the orientation of the plane normal with respect to each interfacing grain (Dechamps, 1991). (b,c) Orientation of GB planes (not the normal to the plane in this case) in nickel with respect to the specimen axes. The non-CSLs (b) are plotted separately to the CSLs (c) (Randle and Dingley, 1989).

6.3 RODRIGUES-FRANK SPACE

The starting point for the representation of any rotation in Rodrigues-Frank space is the angle/axis formulation of the rotation matrix which is used to derive the Rodrigues vector. Although the Rodrigues vector was proposed in the last century, it has only been since the mid 1980s that it has been used as a means to represent rotations (Frank, 1988a,b). Initially RF space was developed as a method for displaying orientation distributions which does not suffer from the inherent disadvantages associated with both Euler space and stereogram-based space (Neumann, 1991). Furthermore, the development of RF space has been in parallel with that for microtexture measurement in polycrystals, particularly by EBSD. This is because the Rodrigues vector (***R***-vector) approach cannot be readily applied to continuous, X-ray generated texture data, but rather is directly applicable to discretely measured orientations or misorientations, i.e. microtexture based measurements. Following this very brief background to the ***R***-vector and RF space, in the remainder of this section we will consider RF space only in the context of the representation of GB misorientations, and not grain orientations.

The Rodrigues vector is defined as

$$UVW\tan(\Theta/2)=\boldsymbol{R} \tag{6.2}$$

where *UVW* is expressed as direction cosines. Hence ***R*** has three components, R_1,R_2,R_3. As an example, the angle/axis 38.94°/110 corresponds to the ***R***-vector with components 1/4,1/4,0. From the definition of the ***R***-vector given in equation 6.2 it follows that the *direction* of the vector specifies *UVW* and its *magnitude* specifies Θ. This is comparable to the definition of the reference sphere misorientation space described in the previous section where a vector is parallel to *UVW* and its length equals Θ.

If it is always the disorientation which is chosen from the 24 symmetry related solutions, then all possible ***R***-vectors for the holosymmetric (i.e. highest symmetry) cubic system will lie in a space bounded by a truncated cube known as the 'fundamental zone' of RF space. The surface of the fundamental zone consists of six regular octagons and eight equilateral triangles as shown in figure 6.8a. Three orthogonal axes, ***XYZ***, which for misorientation representation are parallel to the 100, 010, 001 crystal axes respectively, have their origin at the centre of the fundamental zone. The whole of Rodrigues-Frank space is twenty-four times the volume of the fundamental zone because the complete space encompasses every symmetry-related solution. However, it is generally only necessary to consider misorientation-related ***R***-vectors within the fundamental zone.

If the misorientation axis is expressed with *UVW* all positive, $U > V > W$ and no handedness associated with the misorientation angle, then we do not need to use the whole of the fundamental zone to display misorientations, but only 1/48 of it (Randle, 1990a; Freye et al, 1991). This compares directly to the use of a single unit triangle (i.e. 1/48 of the reference sphere) rather than the whole stereogram in stereogram-based space (Warrington and Boon, 1975). The form of 1/48 of the fundamental zone, which will be called the 'subvolume', can be appreciated from figures 6.8. Figure 6.8a is the whole fundamental zone and figure 6.8b is 1/8 of the fundamental zone. Marked on figure 6.8b is a 'double subvolume' (i.e. a right-handed and left-handed subvolume). The double subvolume is a square-based skew pyramid which has the vertex truncated asymmetrically. If this polyhedron is bisected through its mirror plane we finally obtain the subvolume (figure 6.8c).

It is a unique property of RF space that the locus of misorientation angles about the same misorientation axis is a straight line through the origin (Frank, 1988a,b). The loci of all ***R***-vectors which represent misorientations about the three lowest index axes 100,110 and 111 are indicated on figures 6.8b and c and can be seen to lie along edges of the subvolume. Misorientations about 111 have an R-vector which points towards the centre of an equilateral triangle on the surface of the fundamental zone. The ***R***-vector which lies at the centre of this triangle, labelled A on figures 6.8b and c, is the largest possible disorientation on 111, namely 60°. This is the $\Sigma = 3$ CSL or twin. The angle/axis parameters of all the 'corners' of the subvolume, labelled ABCDE on figures 6.8b and c, are

A - 60.0°/111
B - 62.8°/1,1,$2^{1/2}$-1 (maximum possible disorientation angle)
C - 56.9°/$2^{1/2}$,1,1
D - 60.7°/110
E - 45.0°/100

Grain boundaries with 'special' geometries such as CADs, LAs and CSLs are clearly and simply represented in RF space, allowing any 'mesotexture' (i.e. misorientation texture) in the data to be appraised. A propensity for CAD GBs can be recognised readily in a sample population because there will be a bias of data points along the CAD axis. Low angle GBs are readily apparent because the magnitude of ***R*** is small and so these vectors cluster around the origin. The positions of the ***R***-vectors for CSLs in the subvolume turn out to be particularly interesting and also easy to locate. Table 6.1 shows the components of ***R***-vectors for CSLs with Σ up to 45, listed according to misorientation axis. We see that the components of each R-vector are all

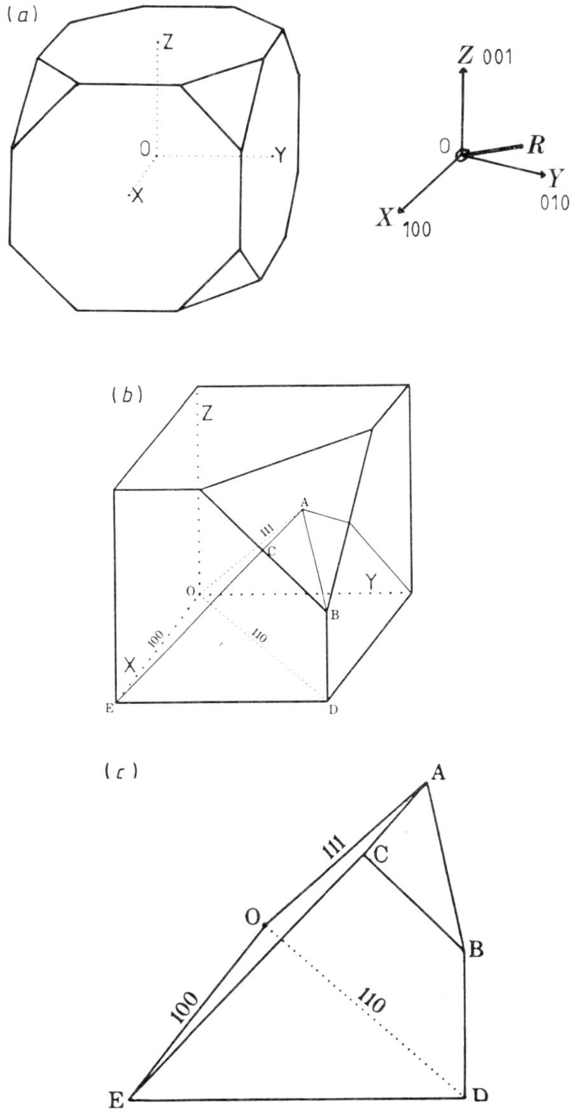

Figure 6.8 Division of the fundamental zone of Rodrigues-Frank (RF) space into 48 basic unit subvolumes, analogous to unit triangles of the stereogram. (a) The whole fundamental zone with the directions of the ***XYZ*** axes indicated and an ***R***-vector (b) 1/8 of the fundamental zone showing a left- and right-handed subvolume. The origin, which is at the centre of the fundamental zone, is labelled O. (c) A single subvolume. The location of the 100, 110 and 111 axes are shown in (b) and (c) (Randle, 1990a).

rational fractions with the misorientation axis given by the numerator. Figures 6.9a and b show the location of CSLs in the subvolume; for clarity only Σ-values up to 35 are included and CSLs misoriented about 112 and 113 are shown separately on figure 6.9b, where the subvolume has been rotated anticlockwise by 90° with respect to figure 6.9a.

TABLE 6.1

COMPONENTS $R_1R_2R_3$ OF THE RODRIGUES VECTOR FOR CSLs, GROUPED ACCORDING TO THE AXIS OF MISORIENTATION FOR THE LOWEST ANGLE SOLUTION, WITH Σ ≤ 45.

Σ	R_1	R_2	R_3	Σ	R_1	R_2	R_3	Σ	R_1	R_2	R_3
100				110				111			
5	1/3	0	0	9	1/4	1/4	0	3	1/3	1/3	1/3
13a	1/5	0	0	11	1/3	1/3	0	7	1/5	1/5	1/5
17a	1/4	0	0	19a	1/6	1/6	0	13b	1/7	1/7	1/7
25a	1/7	0	0	27a	1/5	1/5	0	19b	1/4	1/4	1/4
29a	2/5	0	0	33a	1/8	1/8	0	21a	1/9	1/9	1/9
37a	1/6	0	0	33c	2/5	2/5	0	31a	1/11	1/11	1/11
41a	1/9	0	0	41c	3/8	3/8	0	37c	3/11	3/11	3/11
								39a	1/6	1/6	1/6
								43a	1/13	1/13	1/13
210				211				221			
15	2/5	1/5	0	21b	2/6	1/6	1/6	17b	2/5	2/5	1/5
27b	2/7	1/7	0	31b	2/5	1/5	1/5	29b	2/7	2/7	1/7
41b	2/6	1/6	0	35a	2/8	1/8	1/8	45b	2/9	2/9	1/9
43b	2/9	1/9	0					45c	2/6	2/6	1/6
310				311				331			
37b	3/8	1/8	0	23	3/9	1/9	1/9	25b	3/9	3/9	1/9
				33b	3/11	1/11	1/11	35b	3/11	3/11	1/11
321								332			
39b	3/8	2/8	1/8					43c	3/8	3/8	2/8

Misorientation axes having indices uv0, uvv and uuw lie in the base DEO and the other two 'inner' faces ACEO and ABDO of the subvolume respectively. Axes of the form uv0 lie in the face which is bounded along two edges by 100 and 111. Similarly uuw and uvv axes lie in planes which are partially bordered by 110 and 111, 100 and 111 respectively. The two remaining faces of the subvolume ABC and BCDE form part of the boundary between the fundamental zone and the next zone. Only $\Sigma = 3$ and $\Sigma = 17b$ lie on an 'outer' face of the fundamental zone, which means that they represent the largest possible angular rotation on a particular axis (the disorientation for the $\Sigma = 17b$ CSL is 61.93°/221). All of the descriptions of misorientations contained in this paragraph are apparent from a study of figures 6.8b,c and 6.9a,b.

Figure 6.9c shows the location of CSLs when the whole subvolume is projected onto its triangular face in the *XY* plane, i.e. ODE in figure 6.8b, with OE horizontal. This choice is convenient to the eye, since the projection is now drawn with the origin conventionally sited at the lower left-hand corner. Several CSLs project onto the same point in the *XY* plane; to differentiate between them open circles of various sizes have been used to indicate the CSLs. The circle diameter is approximately proportional to the value of the R_3 component (see Table 6.1) and the actual ***R***-point is located at the centre of the circle. For a regularly shaped space such as Euler space (usually a cube) or misorientation space (usually a cylinder) a sectionalised display is appropriate for the printed page because all the sections are the same size and shape. Although the subvolume in RF space is an irregular polyhedron it is still convenient to show the data in sections if the surface chosen as the base of the polyhedron (that is, the plane parallel to the sections) is taken to be its most regular side, that is, the isosceles triangle as shown in figure 6.9c. A section in R_3 or a projection of all the data points in the subvolume extends to the whole of the triangle in the *XY* plane.

An example of an experimental data set of misorientations from an austenitic steel is shown in figure 6.10. Figure 6.10a is a projection of all the data, and figures 6.10b-g show sections though the subvolume for R_3 having the following ranges: 0 - 0.05, 0.05 - 0.10, 0.10 - 0.15, 0.15 - 0.20, 0.20 - 0.25, >0.25. The volume of each of the sections is superimposed on the triangular base of the subvolume. The location of each section in R_3 is illustrated on figure 6.10h.

The mesotexture aspects of the dataset in figure 6.10, i.e. low angle (LA) GBs, CSLs and CADs, are marked. A subset of LAs is apparent from the presence of some very small ***R***-vectors about the origin in figures 6.10a and b. It is apparent that $\Sigma = 3$ and $\Sigma = 9$ CSLs are strongly represented, and to a lesser extent some other CSLs, which are labelled. The loci of GBs which are misoriented on 210 and 310 are marked on figure 6.10b because it is apparent that there is some CAD clustering near to these axes.

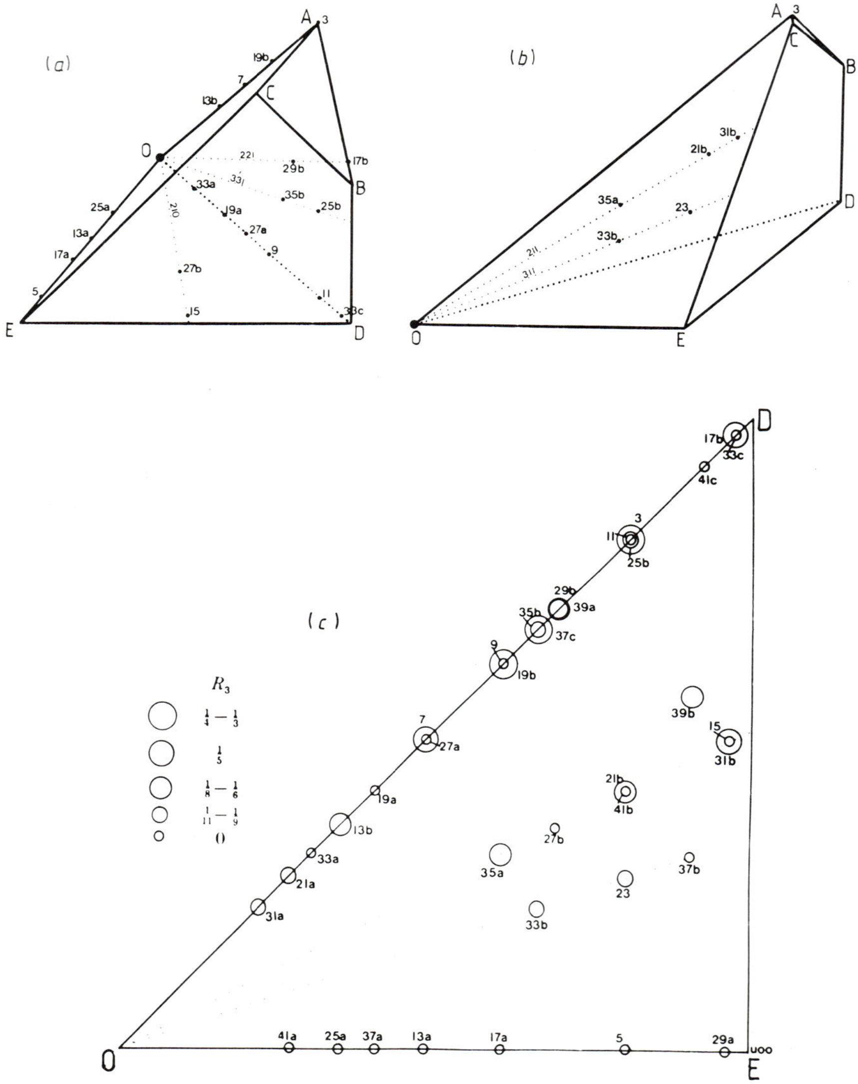

Figure 6.9 Location in the subvolume of ***R***-vectors of CSLs. (a and b) Perspective view of CSLs which are misoriented on (a) 100, 110, 111, 210, 221, 331 and (b) 211, 311 disorientation axes. (b) is rotated by 90° with respect to (a) for greater clarity of presentation. Σ-values up to 35 areincluded. In (c) the ***R***-vectors of CSLs have been projected onto the 'base' of the subvolume, and Σ-values up to 41 are included. Increasing value of R_3 is denoted by increasing size of the open circles which are used to show the ***R***-vectors of the CSLs (Randle 1990a).

Finally, figure 6.10 is the same set of misorientations as in figure 6.3. If these two figures are compared it is apparent that the 'special' geometries are more easily recognised in RF space than in stereogram-based space.

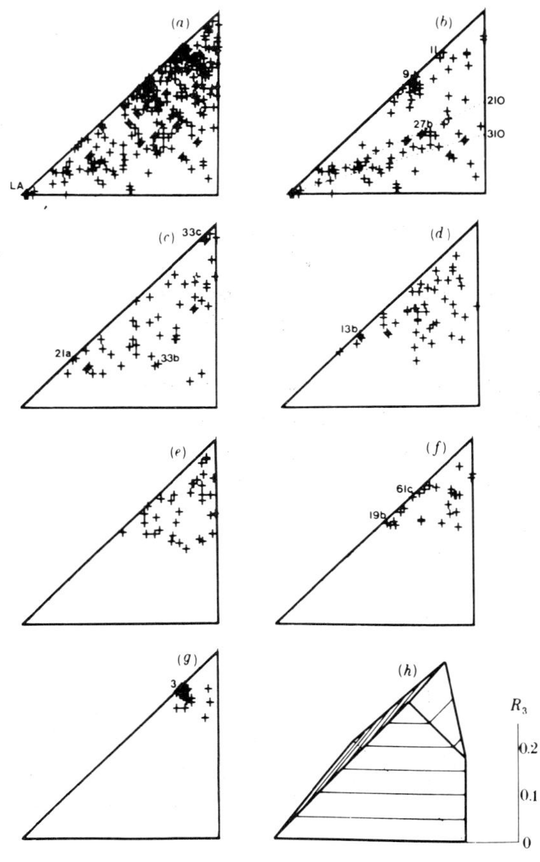

Figure 6.10 Experimentally-generated set of misorientations in RF space. This is the same data that was shown in stereogram-based space in figure 6.3. (a) Projection of all data onto the 'base' of the subvolume. (b-g) Sections in R_3 through the subvolume for the following values: (b) 0-0.05, (c) 0.05-0.10, (d) 0.10-0.15, (e) 0.15-0.20, (f) 0.20-0.25, (g) >0.25. (h) shows how the sections (b-g) are 'stacked' in the subvolume. Clusters of CSLs and LA GBs are marked throughout (Randle, 1990a).

6.4 EULER SPACE

Euler space is the traditional medium for the representation of macrotextures. It is therefore familiar to many workers in the field of texture and this is one reason why Euler space is used for the representation of misorientations in the same manner as it is used for orientations, although RF space offers a more simple and straightforward representation method. Procedures for processing both continuous, X-ray generated macrotexture data and discrete microtexture data to formulate orientation distribution functions (ODFs), have been adapted to produce misorientation distribution functions (MODFs) in Euler space and stereogram-based space (Haessner et al, 1991; Pospiech et al, 1986; Werner and Prantl, 1988). The MODF describes the probability that a GB separates grains of orientation difference (misorientation) δg. This approach becomes a particularly powerful analytical tool when it is combined with the *actual* distribution of GB misorientations, as derived from individual, spatially specific grain orientation measurements. We will examine these concepts in more detail.

The 'physical' MODF (Bunge and Weiland, 1988), also called the 'spatial orientation correlation' (Pledge, 1987) or the 'measured' MODF (Adams et al, 1987; Zhao, Adams and Morris, 1988) is defined as the area fraction of GBs across which the misorientation δg occurs. It is given by

$$F_p(\delta g) = \int \varphi(\delta g, x)\, d\Omega = \int \psi(\delta g, k)\, d\Omega \tag{6.3}$$

$$dA(\delta g)/A = F_p(\delta g)\, d\delta g \tag{6.4}$$

where A is the total area of GBs in the sample population, $F_p(\delta g)$ is the physical MODF, φ and ψ are distribution functions, x and k are the GB normal direction relative to the specimen and crystal systems respectively and $d\Omega$ is the solid angular range increment for x and k.

A distribution which describes the GB geometry in terms of five parameters has been called the 'intercrystalline structure distribution function', ISDF (Adams, 1986). In the ISDF the two parameters which describe the distribution of the GB plane normal directions are also expressed relative to the external, or specimen, reference axes.

In general, if grains are assumed to be randomly distributed in space irrespective of their orientation or shape, then F is given by

$$F_u(\delta g) = \int f(g).f(\delta g.g)\, dg \tag{6.5}$$

where g is the orientation. The distribution in equation 6.5 is the 'uncorrelated MODF', F_u, (Bunge and Weiland, 1988) also sometimes called the 'textural (or statistical) orientation correlation' (Pledge, 1987) or the 'theoretical MODF' (Zhao, Adams and Morris, 1988). An uncorrelated MODF can be obtained from either macrotexture data or microtexture data. The microtexture method is to select at random pairs of measured orientations from a sample population *without regard for their physical positions in the microstructure.* The misorientations of these orientation pairs are then calculated. The uncorrelated MODF thus embodies the probability distribution of misorientations which occur purely on a chance basis.

Although Euler space is the most familiar mode of representation for an MODF, it can also be constructed in 'stereographic space', which is also known as 'rotation parameter space'in the context of MODFs (Pospiech et al, 1986). Here the misorientation is expressed as polar coordinates rather than the Euler angles.

Discrete orientation (and thus misorientation) data can also be plotted directly in Euler space. For the purposes of comparisons, however, they are usually transformed into a continuous distribution function by using a series expansion method which is familiar in texture analysis; this is described elsewhere (Wenk, 1985; Bunge 1985, 1987; Weiland et al, 1988). If the segment lengths of GBs observed in planar cross-sections of the microstructure have been individually measured, these can be used as input to 'weight' each misorientation according to the projected GB length. Figure 6.11a shows an uncorrelated MODF for a sample population taken from an Al-Mn alloy cold rolled 90%. The format of figure 6.11a (and figures 6.11b and c) is that conventionally used to display distributions in Euler space: the three-dimensional cube-shaped space is sliced parallel to one of its square faces into sections of 5°. We see that the identity component (i.e. where all three Euler angles are near zero) is very strong. This can be for two reasons: firstly because there are 24 crystal symmetry variations the identity component appears to be stronger than it really is, and secondly if the orientation population contains a large number of similarly oriented grains (i.e. it is highly textured) then there will be many low angle GBs (i.e. near the identity) in the uncorrelated MODF.

The difference between the physical MODF and the uncorrelated MODF is that the former embodies only the misorientations which actually occur in the microstructure rather than all the statistical possibilities. Figure 6.11b shows the physical MODF obtained from 144 GBs in the Al-Mn alloy. We see that the 'real' misorientation texture in figure 6.11b is less strong than the statistical one in figure 6.11a. This observation can be quantified by defining another function called the 'orientation correlation function', $O(\delta g)$, which is given by

$$O(\delta g) = F_p(\delta g)/F_u(\delta g) \tag{6.6}$$

$O(\delta g)$ acts as a 'filter' to remove the effects on the misorientation distribution of certain GB geometries occuring because of statistical chance in the microtexture. Figure 6.11c shows $O(\delta g)$ obtained as a result of dividing the physical MODF in figure 6.11a by the uncorrelated one in figure 6.11b. This orientation correlation function contains peaks at the origin (i.e. low angle GBs) and at $\varphi_1 = \Phi = 0°$, $\varphi_2 = 45°$. These two GB geometry types are therefore preferred in the microstructure for reasons other than chance proximity. The orientation correlation function should not be confused with the 'orientation coherence function', OCF (Adams et al, 1987; Wang et al, 1990b). This function defines the preference for grains of a specified orientation to reside near other grains of another specified orientation. The 'nearness' is defined by a vector $\boldsymbol{r}$; $\boldsymbol{r}$ separates points p and $'p$ which represent orientations g and $'g$ respectively. Thus the OCF is defined as the probability density for the joint occurrence of orientation g at point p and orientation $'g$ at point $'p$ where p and $'p$ are independently located in a specified measurement volume, and are separated by a vector $\boldsymbol{r}$. For small magnitudes of $\boldsymbol{r}$ the OCF reduces to the MODF, and at large values it describes a local ODF.

Figure 6.12 shows an ISDF for a stainless steel specimen. The GB normal direction is not represented because it was established during this investigation that the GB normal distribution with respect to the specimen was random. Thus effectively the ISDF reduces to a misorientation distribution. Some locations for CSLs which are prevalent in the distribution are marked on figure 6.12, although not all of the peaks correspond to CSLs, for example the peak at $\alpha = 70°$, $\beta = 75°$, $\psi = 20°$ is not a CSL.

It must be noted that the domain of Euler space covered in figure 6.12 is such that each misorientation in the general case (no symmetry) is represented six times by a different symmetry related variant (Van Houtte and Wagner, 1985). This is made clear from figure 6.13 which shows that the fundamental units in Euler space are analogous to stereographic unit triangles. Note that in figure 6.13 the Euler space is sectioned in φ_1 whereas figure 6.12 it is sectioned in β, which is equivalent to Φ. When the axis of misorientation has symmetry properties (i.e. it lies on the edge or corner of the stereographic unit triangle) the number of distinct forms of the misorientation is reduced, which is the situation for CSLs with low Σ-values. Table 6.2 shows 12 of the 24 sets of Euler angles which correspond to the symmetry-related variants of the misorientation for a low-Σ CSL, $\Sigma = 7$. Twelve of these symmetry-related solutions are sufficient to highlight the duplication of Euler angles. Table 6.3 lists the Euler angles of CSLs up to $\Sigma = 27$, calculated from the disorientation matrix, which can be used as a reference for the location of CSLs in Euler space. The multiplicity of Euler space is in contrast to the fundamental zone of RF space where a misorientation is represented by a single point corresponding to the lowest angle variant. Similarly stereogram-based space can be reduced to a single domain, i.e. a single or double unit triangle.

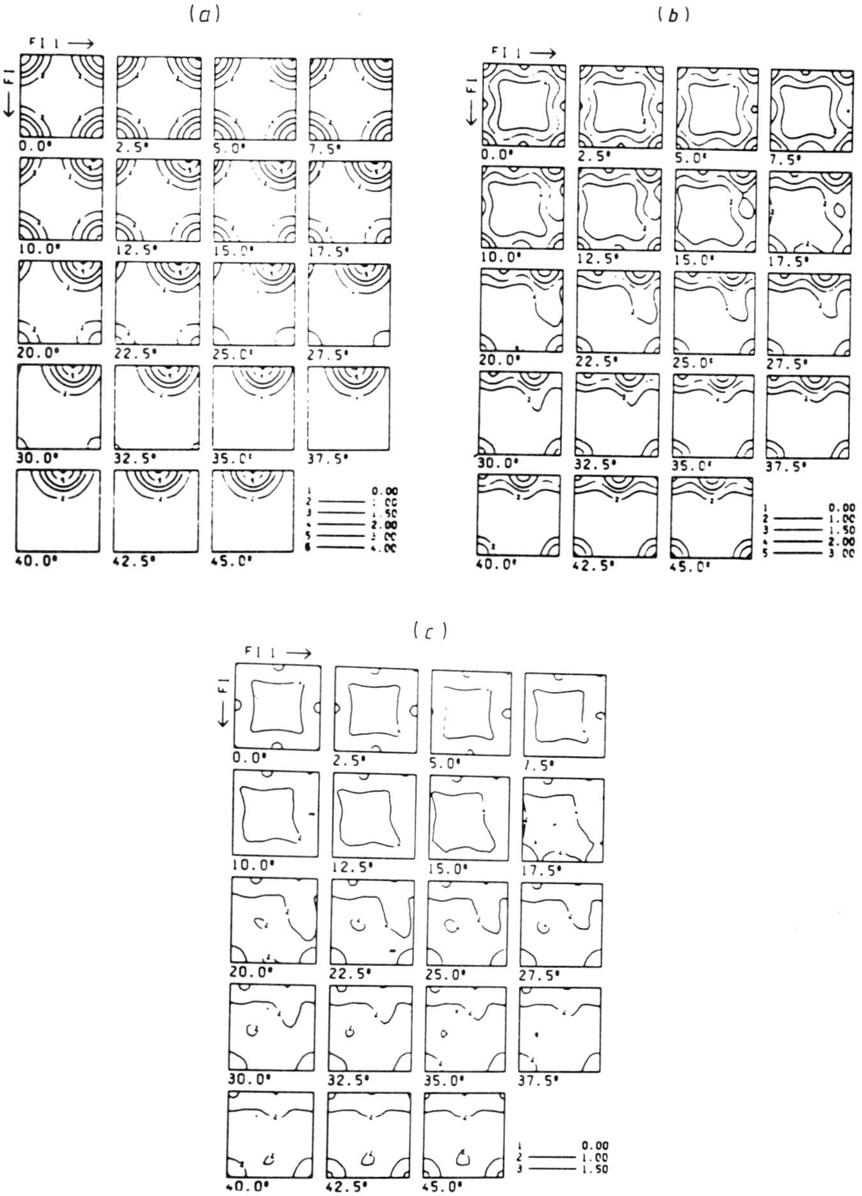

Figure 6.11 Misorientation data from a recrystallised Al-Mn alloy displayed in Euler space. (a) Uncorrelated MODF, derived from the statistical possibilities of misorientation occurrences. (b) Physical MODF, showing the actual misorientation distribution in the microstructure. (c) The orientation correlation function, which is the result obtained when the distribution in (b) is divided by that in (a) (Bunge and Weiland, 1988).

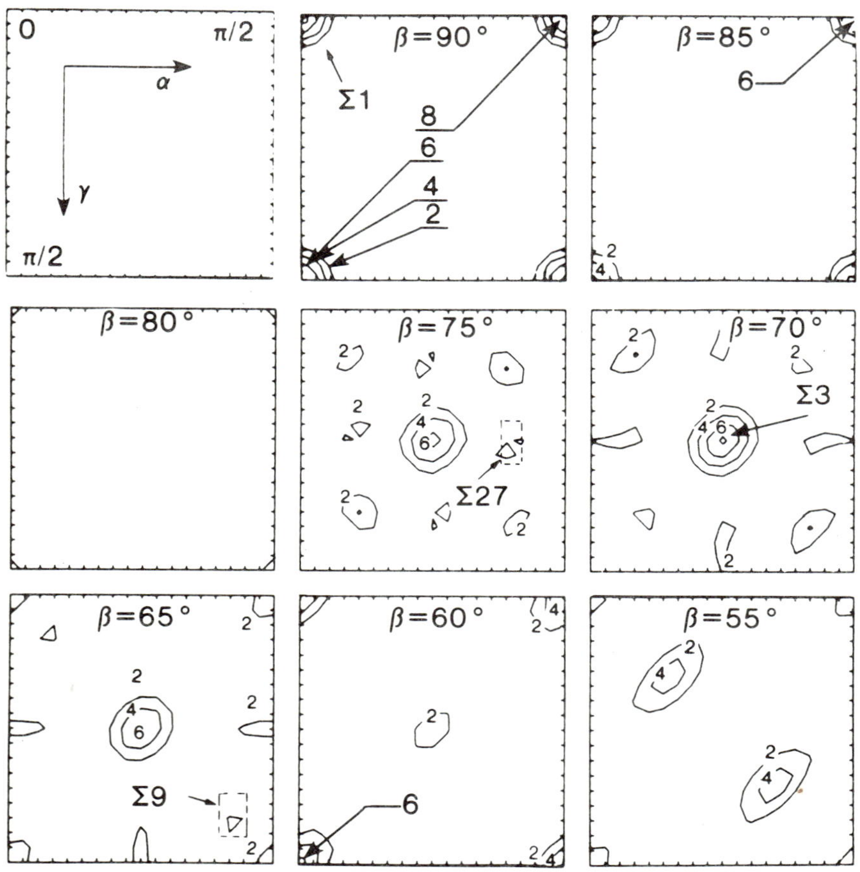

Figure 6.12 Intercrystalline structure distribution function (ISDF) in Euler space for an fcc steel specimen. The GB normal direction is fixed, and the contour levels are in units of times random. CSL peaks are labelled (Zhao, Koontz and Adams 1988).

TABLE 6.2

12 OF THE 24 SYMMETRY-RELATED EULER ANGLES FOR Σ = 7.

φ_1	Φ	φ_2
-33.7	31.0	-56.3
-63.4	73.4	-63.4
33.7	-31.0	-56.3
3.4	-73.4	-63.4
8.4	-64.6	18.4
-33.7	-31.0	-56.3
18.4	64.6	18.4
-56.3	31.0	-56.3
33.7	31.0	-56.3
56.3	31.0	-56.3
26.6	-73.4	-63.4
71.6	64.6	18.4

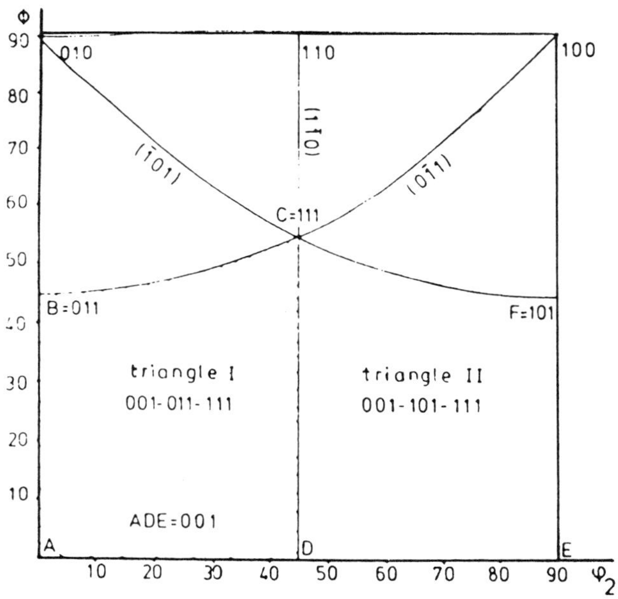

Figure 6.13 Breakdown of Euler space into regions analogous to unit stereographic triangles, illustrating also the distortion of the space (Van Houtte and Wagner, 1985). Note that in this figure the space is sectioned in φ_1, whereas in figure 6.12 it is sectioned in ß, which is equivalent to Φ.

TABLE 6.3

EULER ANGLES FOR CSLS UP TO Σ = 27 CALCULATED FROM THE DISORIENTATION MATRIX

Σ	φ_1	Φ	φ_2
3	153.4	48.2	116.6
5	0	36.9	0
9	135.0	38.9	135.0
11	135.0	50.5	135.0
13a	0	22.6	0
13b	143.1	22.6	126.9
15	153.4	48.2	153.4
17a	0	28.1	0
17b	146.3	58.0	123.7
19a	135.0	26.5	135.0
19b	149.0	37.8	121.0
21a	141.3	17.8	128.7
21b	162.9	40.4	144.0
23	167.9	38.5	155.2
25a	0	16.3	0
25b	141.3	50.2	128.7
27a	135.0	31.6	135.0
27b	153.4	35.4	153.4

The problem of multiple representation of misorientations in Euler space can be solved if a subvolume is defined such that every interior point in it represents a physically distinct misorientation. In other words of all the possible Euler angle representations of a misorientation, one must fall within the subvolume. This subvolume is known as the 'asymmetric domain' of Euler space (Zhao and Adams, 1988; Adams and Zhao, 1990). Its volume and boundaries are defined by the following relations:

$$0 \leq \cos\Phi \leq \sin\varphi_1 \sin\varphi_2/(1+\cos\varphi_1 \cos\varphi_2) \tag{6.7a}$$

$$0 \leq \varphi_1 \leq \varphi_2 \leq \Phi/2 \tag{6.7b}$$

$$\Phi \geq 0 \tag{6.7c}$$

$$\varphi_1+\varphi_2 \leq \Phi/2 \tag{6.7d}$$

The shape and location of the asymmetric domain in Euler space are shown on figure 6.14a. It has the important property that the 'volume trueness' associated with a random distribution is nearly constant. In other words a truly random distribution of misorientations is randomly distributed in the asymmetric domain.

From the definition of the asymmetric domain given above, only one of the 24 variants for a CSL resides in the domain, and this is not usually the one associated with the disorientation. Table 6.4 lists the Euler angle set which falls within the asymmetric domain for Σ up to 49. A former version of this Table contains some errors (Zhao & Adams, 1988). Table 6.4 is the later, corrected version (Adams & Zhao, 1990). Figure 6.14b shows locations of the CSLs listed in Table 6.4 in the asymmetric domain; the subvolume is labelled ABCD which corresponds to the same labelling on figure 6.14a. The coordinates *xyz* of the points in the domain corresponding to CSLs are obtained from the Euler angles as follows:

$$x = (\pi/4)2^{1/2} - 2^{1/2}\varphi_1 - (\varphi_2 - \varphi_1)/2^{1/2} \quad (6.8a)$$

$$y = (\varphi_2 - \varphi_1)/2^{1/2} \quad (6.8b)$$

$$z = \pi/2 - \Phi \quad (6.8c)$$

Except for Σ = 39b, all the CSLs lie on the surface of the domain. Σ = 39b lies within the domain because it is the only CSL in Table 6.4 which has 24 distinct forms for *UVW*. In figure 6.14b we see that CSLs with a disorientation on 100,110 and 111 lie on the DC, CA and AD edges of the domain respectively.

6.5 MISORIENTATION IMAGING/MAPPING

The statistical and spatial aspects of GB geometrical data can be portrayed by a combined representation on an image or diagram of the microstructure (Lorenz and Hougardy, 1988). We could call this 'misorientation imaging' or 'misorientation mapping'. Equally, GB plane information can replace or be included with misorientations. The advantages of this pictorial sort of representation are that the connectivity of GBs in the sample population is made visually apparent, and that other pertinent features of the microstructure such as dihedral GB angles, grain shape, GB precipitation, cavitation or other defect structure are likewise visually appraisable. The disadvantages of an image type of representation is that it is not quantitative and that it is only practicable to show small amounts of data at one time. Examples of

TABLE 6.4

EULER ANGLES FOR CSLs WHICH LIE IN THE ASYMMETRIC DOMAIN OF EULER SPACE

Σ	φ_1	Φ	φ_2	Σ	φ_1	Φ	φ_2
3	45.0	70.5	45.0	33b	12.3	83.0	58.7
5	0	90	36.9	33c	38.7	76.0	38.7
7	26.6	73.4	63.4	35a	16.9	80.1	60.5
9	26.6	83.6	26.6	35b	31.0	88.4	59.0
11	33.7	79.5	33.7	37a	0	90	18.9
13a	0	90	22.6	37b	12.5	85.4	40.6
13b	18.4	76.7	71.6	37c	36.9	71.1	53.1
15	19.7	82.3	42.3	39a	21.8	75.1	68.2
17a	0	90	28.1	39b	29.2	87.1	48.1
17b	45.0	86.6	45.0	41a	0	90	12.7
19a	18.4	87.0	18.4	41b	17.1	84.4	36.0
19b	33.7	71.6	56.3	41c	36.9	77.3	36.9
21a	14.0	79.0	76.0	43a	9.5	82.0	80.5
21b	22.8	79.0	50.9	43b	12.1	87.3	24.8
23	15.3	82.5	52.1	43c	45.0	80.6	45.0
25a	0	90	16.3	45a	10.3	83.6	63.4
25b	36.9	90	36.9	45b	26.6	83.6	63.4
27a	21.8	85.8	21.8	45c	38.7	84.9	51.3
27b	15.1	85.8	31.3	47a	26.6	87.6	63.4
29a	0	90	0	47b	22.7	82.7	35.4
29b	33.7	84.1	56.3	49a	31.0	72.2	59.0
31a	11.3	80.7	78.7	49b	10.6	85.3	47.5
31b	27.4	78.7	43.7	49c	30.4	75.8	49.3
33a	14.0	88.3	14.0				

investigations which use 'misorientation mapping' are listed in Table 7.1.

Figure 6.15 shows an example which typifies the use of a misorientation image. Only a small part of the total data is included in the mapping, which is displayed both in the form of a micrograph showing the location of preferential etching (figure 6.15a) and a schematic diagram of this area showing the geometrical classification of GBs and the location of U-line triple junctions (figure 6.15b) (see section 3.4).

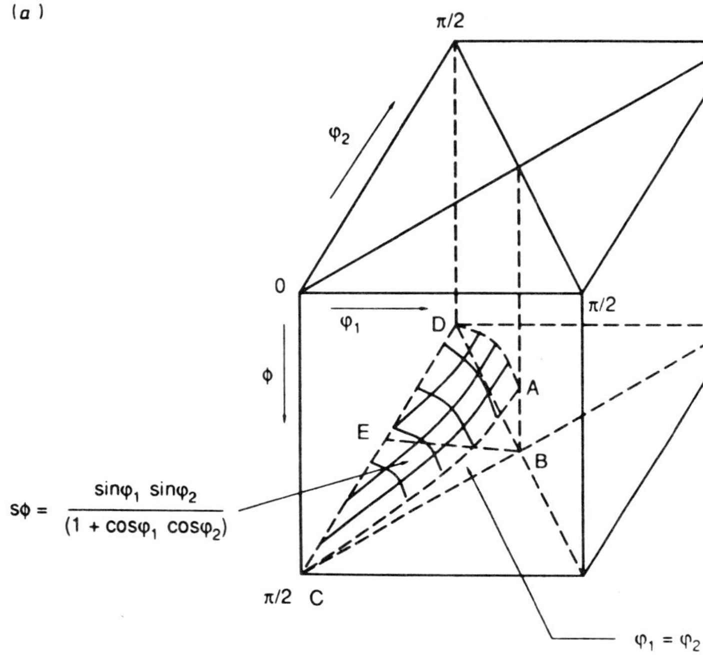

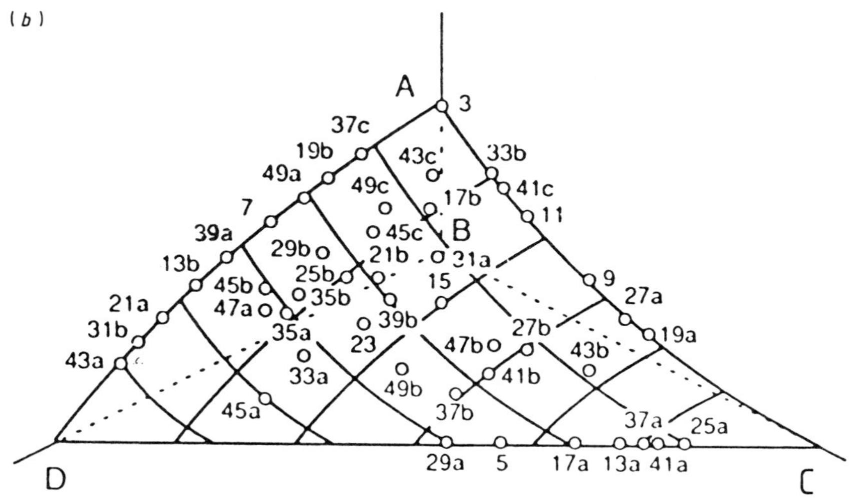

Figure 6.14 The 'asymmetric domain' of Euler space. (a) Location of the domain with respect to Euler space. (b) Positions of CSLs up to $\Sigma = 49$. All CSLs except $\Sigma = 39b$ lie on the surface of the domain. (Zhao and Adams, 1988).

(a)

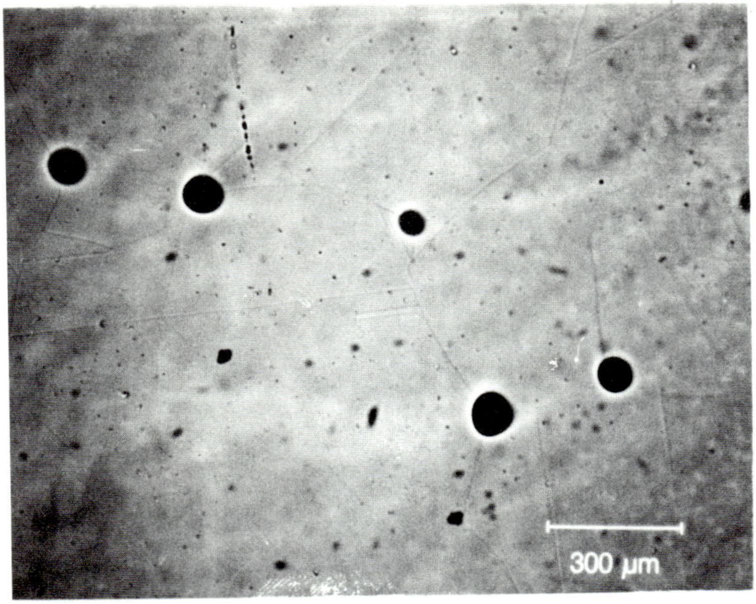

(b)

Figure 6.15 Example of 'misorientation imaging/mapping' in nickel. (a) A micrograph of an analysed region showing corrosion pits at some **U**-lines. (b) Diagrammatic reproduction of (a) giving details of misorientations and triple junction characters. Non-CSL GBs are denoted by R. (Courtesy of G. Palumbo, Palumbo and Aust, 1989).

Reducing the real-space representation of misorientations to one dimension can simplify the display and is especially useful for highlighting cumulative orientations (Orsund et al, 1989) or orientation coherence type behaviour. Figure 6.16 shows a perspective representation of misorientations as Rodrigues vectors along a linear traverse in a deformed aluminium single crystal. This display indicates the wave-like evolution of misorientation with distance.

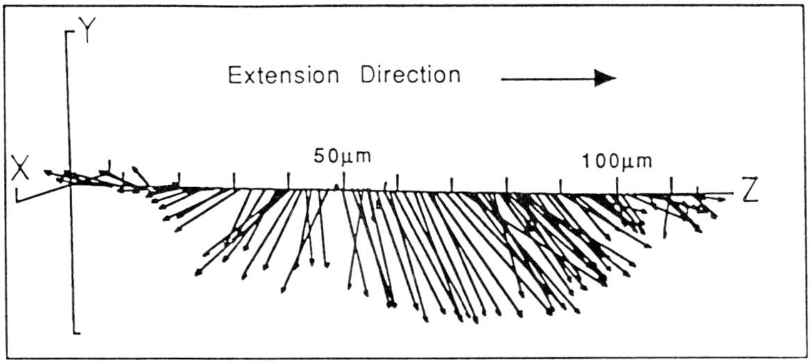

Figure 6.16 Perspective representation by Rodrigues vectors of misorientations along a specimen direction in aluminium. One of the axes in RF space is parallel to the extension direction in the specimen. This representation highlights the cumulative characteristics of the misorientations (Weiland, 1992).

7

EXPERIMENTAL INVESTIGATIONS TO MEASURE GRAIN BOUNDARY GEOMETRY

7.1 INTRODUCTION

In this final chapter we survey how all the topics discussed throughout this book are actually implemented in practice. The survey is in the form of a table, Table 7.1, in which investigations of GB geometry are listed and summarised. For each entry the summary includes the following: the technique used to make the measurements, the materials studied, the topic of investigation, how the data are represented and displayed, the sample population size and, where applicable, the maximum Σ-value for CSL classification. The objective of preparing this survey is to review how GB geometry has been measured and analysed, and how the measurements are evolving with time. The conclusions of each investigation are not the focus of attention in the context of this book, which is essentially concerned with GB geometry measurement as an interpretive tool, and consequently conclusions drawn from experimental data are not included in Table 7.1. However, each entry is referenced and so the full content can be readily traced if required.

In compiling Table 7.1, only those investigations which have been published in the readily-available literature are included. Theses, reports, some conference proceedings and papers published in languages other than English are omitted. Studies on bicrystal specimens are also omitted, because this book is concerned with polycrystals. The lower limit for a polycrystal is a one-dimensional arrangement of grains, i.e. a wire (Omar and Mykura, 1988). Two-dimensional polycrystals exist as thin films or ribbons (Watanabe *et al*, 1989). Both one- and two-dimensional polycrystals are included in Table 7.1.

7.2 SURVEY OF GRAIN BOUNDARY GEOMETRY INVESTIGATIONS

The survey of GB geometrical investigations is divided into four parts, according to which experimental technique has been used to make the measurements - X-ray techniques (principally Laue diffraction), TEM (including HVEM), SAC or EBSD. In Table 7.1 each entry is listed in chronological order under one of these four headings. The Table is subdivided into the following columns:

1. Material(s) investigated. Only a brief description of each material is included; for instance an element symbol may also include very dilute alloys, and steels are labelled generically as fcc or bcc. Further details (e.g. percentages of alloying elements, designatory numbers, etc) are usually available in the original papers.

2. Subject of investigation. Similarly to (1) above, the investigation description is necessarily very condensed to fit into tabular form and reference must be made to the original papers for full details.

3. Output. The abbreviations used for the descriptors of GB geometry are those which have been used throughout this book, i.e. Θ, *UVW*, CSL, *N*, CAD and Euler (angles). A 'prime' symbol after Θ, *UVW* or *N* means that these parameters are measured relative to the specimen axes, rather than the crystal axes. *N*(s) or *N*'(s) means that *N* is measured for only part of the sample population. Where Θ/*UVW* is not explicitly listed, it signifies that specific references were not made, or examples given, of this parameter in the original paper, although Θ/*UVW* values may have been measured. Similarly Θ and *UVW* are sometimes recorded separately or individually in the Table to reflect their usage in the original paper. Low angle GBs are included as CSLs having $\Sigma = 1$, or if the CSL system is not used to categorise the data, as Θ.

4. Display method. The display method is sometimes listed using abbreviations which have been used throughout the book, i.e. MODF (only included if the physical MODF is measured), RF (space), $O(\delta g)$, OCF, ISDF. In addition, the following abbreviations are used to summarise how the GB data is conveyed:

Table - any list which is separate to the text;
Graph - any histogram, chart or plot;
Text - data presented via the text;
Map - any pictorial representation of data which shows spatial relationships of thesampled GBs, e.g. a micrograph with GBs labelled according to their geometry;

PF - 'pole figure', i.e. misorientation axes represented with respect to the specimen axes;
UT-M - misorientation axes represented in a single unit triangle of the stereogram;
UT-MM - as above, with partitioning of the triangle according to the 'Mackenzie triangle';
UT-A - misorientation axes and angles represented in a series of unit triangles according to the angle, or with angles labelled.
UT-P - GB plane normals represented in a single unit triangle of the stererogram.

5. No. GBs, i.e. the number of GBs which comprises the total sample population, where this is given in the original paper. Sometimes the number of grains analysed, rather than the number of GBs, is given in the paper, and this is denoted by 'g' following the number.

6. Σmax, i.e. the maximum Σ-value which is used for categorisation of a GB as a CSL, where the CSL classification is used. Where Σmax is not defined explicitly in the paper, the highest Σ in the experimental data is listed in this column, followed by '+'. For every case except two, the Brandon criterion (equation 3.8) is used as a cut-off for the maximum allowable deviation from a particular CSL, v_m. the two exceptions are:

$$v_m = 7.5° \times \Sigma^{-1/2} \text{ (Mackenzie } et\ al\text{, 1988)} \qquad (7.1a)$$

$$v_m = 15° \times \Sigma^{-5/6} \text{ (Palumbo and Aust, 1990a)} \qquad (7.1b)$$

7. Reference.

7.3 COMMENTS ON TABLE 7.1 AND OUTLOOK FOR THE FUTURE

There are a number of trends in the measurement of GB geometry which are apparent from Table 7.1. With regard to technique usage, Table 7.2 summarises the number of GB geometry investigations conducted using each of the four techniques - Laue, TEM, SAC and EBSD. Up to the present time TEM has been the most widely used technique, often for investigations involving deformation structures or segregation. Table 7.2 also shows the average sample population size of GBs, compiled from such data as is available. There is an order of magnitude difference between the average number of GBs analysed by TEM and SEM-based techniques which clearly

TABLE 7.1

SUMMARY OF EXPERIMENTAL INVESTIGATIONS TO MEASURE GRAIN BOUNDARY GEOMETRY IN POLYCRYSTALS, CLASSIFIED ACCORDING TO TECHNIQUE USED.

X-rays (Laue)

Material	Subject of investigation	Output	Display method	No. GBs	Σmax	Reference
**Al	Nucleation of recrystallisation	Θ/*UVW* PF	Table Map	-	-	Bellier & Doherty 1977
Ni	GB cavitation	Θ/*UVW* CSL	Graph	-	75	Lim & Raj 1984a
Ni	GB statistics	Θ/*UVW* CSL	Table Graph	222	97	Lim & Raj 1984b
Ni alloy	GB diffusivity	Θ/*UVW* CSL	UT-A Graph	37	17	Funkenbusch & Giamei 1986
Al	Recrystallisation GB dissociation	CSL Text	Map	-	27+	Kopezky & Fionova 1990
Cu	Recrystallisation of single crystal	Euler Table	MODF	214	101	Haessner *et al* 1991
CuP CuMn	Recrystallisation of single crystal	Θ/*UVW*	RF	-	-	Freye *et al* 1991

TEM

Material	Subject of investigation	Output	Display method	No. GBs	Σmax	Reference
fcc steel	Discontinuous precipitation	Θ/*UVW* CSL	Text N(s)	-	55	Ainsley *et al* 1979

Table 7.1 (*continued*)

Al	GB sliding & dislocations	CSL	Graph Table	-	25+	Kokawa *et al* 1981
bcc Fe-M-C	GB dislocations & precipitation	Θ/*UVW* CSL CAD	Table	-	17	Lartigue & Priester 1983
Cu	Deformation characteristics	Θ' *UVW*'	Map PF UT-M MODF	205g	-	Haessner *et al* 1983
*Al	Recrystallisation of single crystal	Θ/*UVW* CSL	Table Map	-	-	Berger *et al* 1983
Al + Al_2O_3	GB pinning Grain growth	Θ/*UVW* CSL	Table	32	101	Tweed *et al* 1984
NiO	GB statistics	CSL *N*	Graph UT-P	71	41	Dechamps *et al* 1985
bcc steel	GB statistics	Θ/*UVW* CSL	Table UT-M	50	49	Martikainen & Lindroos 1985
Ni	S segregation	CSL CAD *N*	Graph Table	100	19	Bouchet & Priester 1986
*fcc steel	Creep cavitation	Θ/*UVW* CSL *N*(s)	Graph Table	-	49	Don & Majumdar 1986
Ni	S segregation	CSL *N*	Table 1987	30	19	Bouchet & Priester
NiO	GB statistics & diffusion	CSL *N* UT-P	Graph Table	100+	41	Marrouche *et al* 1987

Table 7.1 (*continued*)

Ni alloy	GB pinning	Θ/*UVW* CSL	Table	200+	49	Randle & Ralph 1987a
*Al & Cu+M	Early stages of recrystallisation	Θ/*UVW* CSL	Table Maps	-	-	Berger *et al* 1988
Al-Mn	Orientation correlations	Euler O(δg)	MODF	144	-	Bunge & Weiland 1988
Fe-P	GB segregation	CSL *N*	Graph UT-P	50	75	El M'Rabat & Priester 1988
α/ß brass	GB statistics X^2 test	Θ *UVW*	UT-MM Graph	-	-	Prantl *et al* 1988
dual phase steel	GB statistics	Euler	MODF	100α	-	Schwarzer & Weiland 1988
+Cu, Al steels Ni alloy	GB statistics	CSL CAD UT-M	Graph Map	182	49	Randle & Ralph 1988c
fcc steel	GB statistics Heat treatment	CSL *N*	Table	70	33	Randle 1989
fcc steel	Recrystallisation	CSL Graph	UT-M	163	35	Randle 1990c
Cu	GB cavitation fatigue	CSL	Graph Table	95	-	Butron-Guillen *et al* 1990
Al-Mg-Zr	Recrystallisation Superplasticity	Θ CSL	Graph	150	29	Hales *et al* 1990
NiO	GB planes/surface Grain growth	CSL *N*	Graph UT-P	250	25	Dechamps 1991

Table 7.1 (*continued*)

fcc steel	GB segregation of Cr	Θ/*UVW* CSL *N*	Table	50	49	Laws & Goodhew 1991
fcc steel	GB statistics SFE	Θ/*UVW* CSL	Graph UT-M	341 +3[6]	29	Gertsman & Tangri 1991
Ni	Early stages of recrystallisation	Θ/*UVW* CSL	Table	85	33	Randle 1990b
Al	Subgrain misorientation	Euler RF Graph	MCI	550	-	Weiland *et al* 1991
*CuMn	Nucleation of recrystallisation	Θ/*UVW*	Text	-	-	Klement *et al* 1991
Cu	Recrystallisation	Θ/*UVW* Euler	MODF Text	470	45+	Sztwiertnia & Haessner 1991

SAC

Material	Subject of investigation	Output	Display method	No. GBs	Σmax	Reference
Fe-P	GB segregation of P	*N*	Graph UT-P	-	-	Suzuki *et al* 1981
Cu-Ag sintered	Dihedral angles	CSL	Graph	-	19	Kaysser *et al* 1982
Al α-iron ß-brass	Intergranular fracture	CSL	Graph Map	-	29	Watanabe 1984
fcc steel	Orientation coherence	Euler Table	OCF	1762g	-	Adams *et al* 1987

Table 7.1 (*continued*)

Fe-3%Si	Secondary recrystallisation	CSL $N'(s)$	Graph Table	1000	33	Harase *et al* 1987
bcc Fe-Ni-Cr	GB segregation of P	CSL $N(s)$	Graph UT-P	292g 350	25	Ogura *et al* 1987
fcc steel	Orientation coherence	Euler Text	MODF 4000	1762g	-	Zhao, Adams, Morris 1988
fcc steel	GB statistics	Euler N'	ISDF MODF	-	27+	Zhao, Koontz, Adams 1988
Cu wire	GB energy	CSL	Graph	-	45	Omar & Mykura 1988
Ni	Secondary recrystallisation	CSL Map	PF	100g	-	Makita *et al* 1988, 1990
Ni	Triple-line corrosion	CSL	Map	-	49	Palumbo & Aust 1989
+ Fe-3%Si	Secondary recrystallisation	CSL	Table	-	-	Rouag & Penelle 1989
Fe-3%Si	Secondary recrystallisation	CSL Graph	Map	-	51	Shimizu & Harase 1989
Al	Fatigue GB migration	Θ/UVW CSL	Table Map	150	29	Raman *et al* 1989
Fe-Si ribbon	GB statistics Texture	CSL	Table Graph	488	29	Watanabe, Arai *et al*, 1989; Watanabe 1990
Fe-Si ribbon	Ductility	CSL Table	Graph	520	29	Watanabe, Fujii *et al* 1989

Table 7.1 (*continued*)

Fe-3%Si	Secondary recrystallisation	CSL	Graph	33		Shimizu *et al* 1990
Ni	GB corrosion	CSL	Graph	-	49	Palumbo & Aust 1990a
Fe-Co	Magnetic anneal Recrystallisation	CSL Table	Graph	-	29	Watanabe *et al* 1990
Ni-S	GB statistics S segregation	CSL Graph Map	Table	-	49	Palumbo & Aust 1990b
+ Fe-3%Si	GB statistics Grain growth	CSL CAD	Table	200	25	Rouag *et al* 1990
bcc/fcc iron	Phase transformation	CSL Graph Map	UT-M	-	45	Harase *et al* 1990
bcc steel	Mechanism of recrystallisation	Θ *UVW* Map	Graph UT-A	1621	-	Plutka & Hougardy 1991
fcc steel	GB statistics	*UVW* CSL UT-M	Graph Table	1897	49	Crawford & Was 1991
fcc steel	Stress corrosion cracking	CSL	Table Graph	2155	49	Crawford & Was 1992
fcc steel	Stress corrosion cracking	CSL	Table	51	49	Palumbo *et al* 1992

EBSD

Material	Subject of investigation	Output	Display method	No. GBs	Σmax	Reference
Ni_3Al	GB statistics Ductile/brittle	CSL Table	Graph 137	400	29	Farkas *et al* 1988

Table 7.1 (*continued*)

Ni_3Al	GB statistics B segregation	CSL	Graph	756	19	Mackenzie *et al* 1988
fcc steel	Strain-induced grain growth	*UVW* CSL	UT-MM Table	300g	37	Randle & Brown 1988
Ni	Grain growth	*UVW* CSL Table	UT-MM Graph	-	49	Randle & Ralph 1988a
Al	Deformation	Θ PF	Graph	-	-	Orsund *et al* 1989
fcc steel	Secondary recrystallisation	*UVW* CSL Graph	UT-M Table	461g	31	Randle & Brown 1989
fcc	GB sensitisation	CSL	Graph	-	33	Ortner & Randle 1989
Ni	GB statistics Grain rotations	CSL *N* *N'*	Table UT-P Graph	137, 47	49	Randle & Dingley 1989,1990a
fcc steel	GB statistics RF space	Θ/*UVW* CSL	UT-A RF	-	49	Randle 1990a
Cu	Interface damage	Euler CSL *N'*	ISDF IDF	-	49	Adams, Zhao O'Hara 1990
Cu	GB cavitation Creep	Euler	IDF Map	10000 2100g	-	Field *et al* 1991
Al	Orientation coherence during deformation	Euler	OCF	2441g	-	Wang, Morris, Adams 1990
Al	Nucleation of Recrystallisation	Θ PF	Graph	-	-	Hjelen *et al* 1991

Table 7.1 (*continued*)

Fe- 3%Si	GB statistics Recrystallisation	Euler CSL	MODF	3000g	25+	Penelle *et al* 1991
Ni	GB statistics Twinning	CSL	Map Graph	2400	35	Furley & Randle 1991
Cu	Fatigue crack initiation	Θ/*UVW* CSL *N*	Table	-	49	Liu *et al* 1992
bcc steel	Secondary recrystallisation	*UVW* CSL	UT-MM Table	400	19	Rabet *et al* 1992
Cu	GB cavitation Creep	Euler *N*' Θ	IDF Graph	7000	-	Field & Adams 1992
Ni	GB planes Grain growth	CSL *N*	Graph Table	200	35	Randle 1991c

** - X-ray Kossel in TEM
* - HVEM
+ includes data from other papers

highlights both the labour intensive nature of the TEM route and the limitations of thin-foil specimens. For SEM-based routes in particular, the number of GBs which comprises the sample population is greater than the number of grain orientations measured, depending upon how the sample population of grains is selected from the microstructure. For example, one entry in Table 7.1 refers to 10,000 GBs associated with 2100 grains. The number of large-scale sample populations has increased recently: 7 out of the 8 samples sizes in excess of 1000 GBs are dated 1991 or 1992. Development work is in progress to automate the EBSD procedure which will facilitate further the collection of large sample populations of orientations (Wright and Adams, 1992; Juul Jensen, 1992; Krieger Lassen *et al,* 1992).

The trend for increased sample population sizes emphasises the need for efficient and appropriate methods of statistical analysis, classification and data presentation. 76% of the entries in Table 7.1 analyse data via the CSL model and 18% via Euler angles which are subsequently displayed in Euler space as an MODF or similar representation. These two routes tend to be mutually exclusive: only 4% of investigations employ both Euler space and the CSL model. The most popular means of data display is by use of graphs and/or tables, which are used in almost every investigation. Use of RF space is reported in only 2.5% of cases. Stereogram-based space is used in 27% of cases, particularly where grain boundary planes are involved. Grain boundary plane normals (involving some or all of the data) are measured crystallographically in 19% of cases.

TABLE 7.2

NUMBER OF INVESTIGATIONS AND AVERAGE NUMBER OF GBs INCLUDED FOR EACH TECHNIQUE

Technique	Total number of investigations	Number of investigations with specified no. GBs	Average No. GBs
Laue	6	1	222
TEM	31	22	149
SAC	26	12	1045
EBSD	18	12	2295

There is a large variation in the quoted upper values for Σ, varying from 19 to 101. During the last three years the view on what is an appropriate value for Σ_{max} has tended to stabilise at between 25 and 49. The modal value is 49, being chosen more than twice as often as the other populous values, $\Sigma = 29$ or 25.

Finally, we will consider briefly the applications of GB geometry measurements in Table 7.1. The materials investigated are almost all fcc (e.g. aluminium, nickel, steels), with the exception of Fe-3%Si which is technologically an important bcc alloy for study of secondary recrystallisation. Studies of 'GB statistics' feature frequently in Table 7.1. Essentially this refers to assessments of the proportions of 'geometrically special' GBs in a specimen, and it may be the sole purpose of the investigation or part of a broader scope. Investigations concerning aspects of recrystallisation are common, particularly the early stages of primary recrystallisation.

Over the next few years the trend for large data sets, i.e. of the order of thousands of GBs, will undoubtably continue, thus giving improved statistical reliability. The increased numbers pose challenges for data representation, for instance a greater requirement to display several parameters at a time. To these ends RF space may be adopted more widely to display misorientations, and more sophisticated forms of GB mapping may be developed, with both of these including the use of colour. The number of papers which report use of EBSD to measure GB parameters is currently less than for TEM or SAC since EBSD is relatively new compared to TEM or SEM and the first EBSD GB papers were only published in 1988. It is likely that a greater proportion of GB geometry investigation will use EBSD in the next few years as the technique becomes more widespread. However, it is also clear from the nature of the topics investigated using TEM that TEM will remain the optimum technique for some applications. GB plane normals can be measured in both the TEM and SEM and, with increased understanding of the role of this parameter in linking GB geometry to properties, it is likely that their measurement will begin to feature more commonly in GB geometry studies. Furthermore, as the understanding of the factors which control GB geometry and properties is increased, it is already becoming evident that this area of materials technology will have increasing relevance to commercially important materials.

APPENDIX

RULES FOR RIGHT-ANGLED SPHERICAL TRIANGLES

A spherical triangle is defined as having sides bounded by the intersection of three great circles on a reference sphere; thus the *sides* of a spherical triangle are arcs of great circles (a, b, c on figure A.1). The intersections of the sides of the triangle give the *angles* (A, B, C on figure A.1). Clearly both a, b, c and A, B, C are measured in degrees or radians. There are fundamental relationships which relate the angles and sides of a general spherical triangle, analogous to the sine and cosine rule in conventional trigonometry.

Where A, B or C is a right-angle, i.e. formed by the intersection of two mutually perpendicular zones, solution of the triangle (referred to as a Naperian triangle is simplified, and the relationships between sides and angles are:

$$\sin a = \tan b \cdot \cot B \tag{A.1}$$
$$\cos B = \tan a \cdot \cot c \tag{A.2}$$
$$\cos c = \cot B \cdot \cot A \tag{A.3}$$
$$\cos A = \cot c \cdot \tan b \tag{A.4}$$
$$\sin b = \cot A \cdot \tan a \tag{A.5}$$
$$\sin a = \sin A \cdot \sin c \tag{A.6}$$
$$\cos B = \cos b \cdot \sin A \tag{A.7}$$
$$\cos c = \cos a \cdot \cos b \tag{A.8}$$
$$\cos A = \sin B \cdot \cos a \tag{A.9}$$
$$\sin b = \sin c \cdot \sin B \tag{A.10}$$

where C is the right-angle as shown in figure A.1. Thus if a Naperian triangle has two known components, the triangle can be solved. Similar rules to those in equations A.1 - A.10 can be formulated for right-sided Naperian triangles, although these are used less frequently. Further information about spherical trigonometry can be found elsewhere (Mckie and Mckie, 1974).

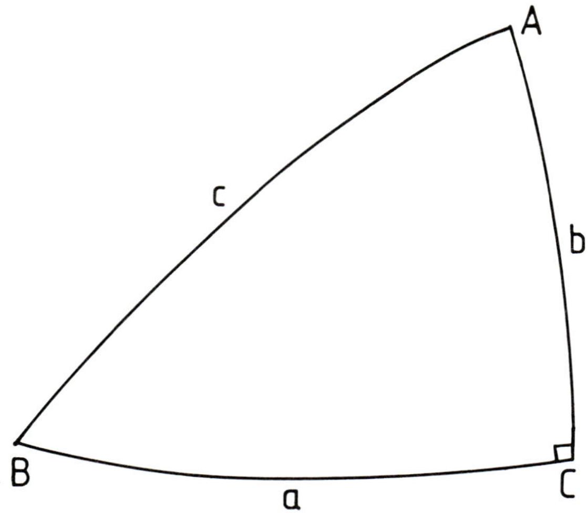

Figure A1 A spherical triangle with sides a,b,c and angles A,B,C.

REFERENCES

A.F. Acton and M. Bevis, Acta Cryst., **A27** (1971) 175.
B.L. Adams, Met. Trans., **17A** (1986) 2199.
B.L. Adams, P.R. Morris, T.T.Wang, K.S. Willden and S.I. Wright, Acta Met., **35**, (1987) 2935.
B.L. Adams and J. Zhao, Acta Cryst., **A46** (1990) 620.
B.L. Adams, J.W. Zhao and D. O'Hara, Acta Met. Mat., **38** (1990) 953.
M.H. Ainsley, G.J. Cocks and D.R. Miller, Met. Sci., **13** (1979) 20.
S.M. Allen and E.L. Hall, Phil. Mag., **46A** (1982) 243.
A.V. Andrejeva, G.I. Salnikov and L.K. Fionova, Acta Met., **26** (1978) 1331.
M.F. Ashby, F. Spaepen and S. Williams, Acta Met., **26** (1978) 1674.
K.T. Aust and G. Palumbo, 'Structure and Property Relationships', Ed. A.H. King and K. Tangri, American Society for Metals, (1991) 3.
K.T. Aust and J.W. Rutter, Trans. TMS-AIME, **215** (1959) 119.
K. Baba-Kishi, PhD Thesis, University of Bristol, UK, (1986).
K. Baba-Kishi and D.J. Dingley, J. App. Cryst., **22** (1989) 89.
R.W. Balluffi, Met. Trans., **13A** (1982) 2070.
R.W. Balluffi and T.Y. Tan, Scripta Met., **6** (1972) 1033.
R.W. Balluffi, M. Ruhle and A.P. Sutton, Mat. Sci. Eng., **89** (1987) 1.
R. Becker and S. Panchanadeeswaran, Tex. Mic., **10** (1989) 167.
S.P. Bellier and R.D. Doherty, Acta Met., **25** (1977) 521.
A. Berger, P.J. Wildbrandt and P. Haasen, Acta Met., **31** (1983) 1433.
A. Berger, P.J. Wilbrandt, F. Ernst, U. Klement and P. Haasen, Prog. Mat. Sci., **32** (1988) 1.
G.L. Bleris, G. Nouet, S. Hagege and P. Delavignette, Acta Cryst., **A38** (1982) 550.
W. Bollmann, 'Crystal Lattices, Interfaces, Matrices', W. Bollmann, Geneva, Switzerland. (1982). W. Bollmann, Phil. Mag., **49A** (1984) 73.
W. Bollmann, Mat. Sci. Eng., **A113** (1989) 129.
R. Bonnet, E.Cousineau and D.H. Warrington Acta Cryst., **A37** (1981) 184.
D. Bouchet and L. Priester, Scripta Met., **20** (1986) 961.
D. Bouchet and L. Priester, Scripta Met., **21** (1987) 475.
D.G. Brandon, Acta Met., **14** (1966) 1479.
D.G. Brandon, B. Ralph, S. Ranganathan and M.S. Wald, Acta Met., **12** (1964) 813.
H.J. Bunge in 'Preferred Orientation in Deformed Metals and Rocks: an Introduction to Modern Texture Analysis', Ed. H.-R. Wenk, Academic Press Inc., UK, (1985) 73.

H.J. Bunge, Int. Met. Rev., **32** (1987) 265.

H.J. Bunge and H. Weiland, Tex. Mic., **7** (1988) 231.

M.P. Butron-Guillen, J.G. Cabanas-Moreno and J.R. Weertman, Scripta Met. Mat., **24** (1990) 991.

C.B. Carter, Acta Met., **36** (1988) 2753.

G.A. Chadwick and D.A. Smith (Eds.), 'Grain Boundary Structure and Properties', Academic Press, New York, USA (1976).

D. Chescoe and P.J. Goodhew, 'The Operation of Transmission and Scanning Electron Microscopes', Royal Microscopical Society handbook no. 20, Oxford Science Publications, Oxford, UK, (1990).

J.W.Christian, 'The Theory of Transformations in Metals and Alloys', Pergamon Press, Oxford, UK, (1965).

D.C. Crawford and G.S. Was, J. Elect. Micros. Tech., **19** (1991) 345.

D.C. Crawford and G.S. Was, Met. Trans. A, (1992), in press.

M. Dechamps, Proc. ICOTOM9, Tex. and Mic., **14-18** (1991) 733.

M. Dechamps, F. Barbier and A. Marrouche, Acta Met., **35** (1987) 101.

M. Dechamps, A. Marrouche, F. Barbier and A. Revcolevschi, J. de Phys., **46** (1985) C4-435.

M. Dechamps, S. Yazidi and F. Barbier, Science of Ceramics **14**, Institute of Ceramics, Stoke-on-Trent, UK, (1988) 569.

G. Dimou and K.T. Aust, Acta Met., **22** (1974) 27.

D.J. Dingley, Scan. Elect. Micros., **IV** (1981) 273.

D.J. Dingley, Inst. Phys. Conf. Ser. No. 119, Institute of Physics Publishing Ltd., Bristol, UK, (1991) 551.

D.J. Dingley and V. Randle, J. Mat. Sci., in press (1992).

D.J. Dingley, K. Baba-Kishi and V. Randle, 'An Atlas of Electron Back-scatter Diffraction Patterns', Institute of Physics Publishing, Bristol, UK, (1992) in press.

J. Don and S. Majumdar, Acta Met., **34** (1986) 961.

E.G. Doni and G.L. Bleris, Phys. Stat. Sol., **110A** (1983) 383.

R.C. Ecob, J. Mic., **137** (1985) 3.

R.F. Egerton, Ultramicros., **6** (1981) 297.

B. El M'Rabat and L. Priester, Mat. Sci. Eng., **A101** (1988) 117.

D. Farkas, M.O. Lewis and V. Rangarajan, Scripta Met., **22** (1988) 1195.

D.P. Field and B.L. Adams, Acta Met. Mat. **40** (1992) 1145.

D.P. Field, B.L. Adams and J. Zhao, Proc. ICOTOM9, Tex. and Mic., **14-18** (1991) 977.H.F. Fischmeister, J. de Phys., **46** (1985) C4-3.

C.T. Forwood and L.M. Clarebrough, 'Electron Microscopy of Interfaces in Metals and Alloys', Institute of Physics Publishing, Bristol, UK, (1991).

F.C. Frank, Met. Trans., **19A** (1988a) 403.

F.C. Frank, Mat. Res. Soc. Bull., March (1988b) 24.

D. Freye, C. Funke, U. Klement and P.-J. Wilbrandt, Proc. ICOTOM9, Tex. and Mic., **14-18** (1991) 1185.

A.W. Funkenbusch and A.F. Giamei, Proc. 'Interface Migration and Control of Microstructure', Ed. C.S. Pande et al, American Society for Metals, Ohio, USA, (1986) 161.J. Furley and V. Randle, Mat. Sci. Tech. **7** (1991)

A. Garbacz and M.W. Grabski, Scripta Met., **23** (1989) 1369.

A. Garg, W.A.T. Clark and J.P. Hirth, Phil. Mag., **A59** (1989) 479.

J. Gastaldi, C. Jourdan, G. Grange and C.L. Bauer, Phys. Stat. Sol., **109A** (1988) 403.

V.Y. Gertsman and K. Tangri, Phil. Mag., **64A** (1991) 1319.

H. Gleiter, Mat. Sci. Eng., **52** (1982) 91.

H. Gleiter, J. de Phys., **46** (1985) C4-393.

P.J. Goodhew, Met. Sci., **13** (1979) 108.

P.J. Goodhew, in 'Grain Boundary Structure and Kinetics', Ed. R.W. Balluffi, American Society for Metals, Ohio, USA, (1980) 155.

P.J. Goodhew and D.A. Smith, Scripta Met., **14** (1980) 59.

P.J. Goodhew, T.Y. Tan and R.W. Balluffi, Acta Met., **26** (1978) 557.

G. Gottstein, Scripta Met., **20** (1986) 1791

C. Goux, Can. Metall. Q., **13** (1974) 9.

M.W. Grabski, J. de Phys., **46** (1985) C4-567

H. Grimmer, Acta Cryst., **A45** (1989) 320.

H. Grimmer, W. Bollmann and D.H. Warrington, Acta Cryst., **A30** (1974) 197.

H. Grimmer, R. Bonnet, S. Lartigue and L. Priester, Phil. Mag., **A61** (1990) 493.

C.R.M. Grovenor, H.T.G. Hentzell and D.A. Smith, Acta Met., **32** (1984) 773.

W. Gust, S. Mayer, A. Bogel and B. Predel, J. de Phys., **46** (1985) C4-537.

P. Haasen, Proc. 'Recrystallisation '90', Ed. T. Chandra, Min., Met. Mat. Soc., USA, (1990) 17.

F. Haessner, J. Pospiech and K. Sztwiertnia, Mat. Sci. Eng., **57** (1983) 1.

F. Haessner, K. Sztwiertnia and P.-J Wilbrandt, Tex. and Mic., **13** (1991) 213.

S.J. Hales, T.R. McNelley and R. Crooks, Proc. 'Recrystallisation '90', Ed. T. Chandra, Min. Met. Mat. Soc., USA, (1990) 231.

J. Harase, R. Shimizu and T. Watanabe, Proc. 7th Riso Int. Symp., 'Annealing Processes', Ed. N. Hansen et al, Riso Press, Denmark (1986) 343.

J. Harase, R. Shimizu and N. Takahashi, Proc. 'Recrystallisation '90', Ed. T. Chandra, Min. Met. Mat. Soc., USA, (1990) 459.

J. Harase, R. Shimizu and N. Takahashi, Proc. ICOTOM9, Tex. and Mic., **14-18** (1991) 679.

J. Harase, R. Shimizu, Y. Nakamura and N. Takahashi, Mat. Forum, **14** (1990) 276.

F. Hargreaves and R.J. Hill, J. Inst. Met., **41** (1929) 257.

P. Heilmann, W.A.T. Clark and D.A. Rigney, Ultramicros., **9** (1982) 365.

P.B. Hirsch, A. Howie, R.B. Nicholson, D.W. Pashley and M.J. Whelan, 'Electron Microscopy of Thin Crystals', Butterworths, UK, (1965).

J. Hjelen, Proc. 'Microscale Textures of Materials', Tex. and Mic., in press (1992).

J. Hjelen, R. Orsund and E. Nes, Acta Met. Mat., **39** (1991) 1377.

J. Humphreys, Proc. Eighth International Conference on Textures of Materials, (ICOTOM8), Ed. J.S. Kallend and G. Gottstein, Met. Soc. Inc., USA, (1988) 171.

Y. Ishida and M. McLean, Phil. Mag. **27** (1973) 1125.

O. Johari and G. Thomas, 'The Stereographic Projection and its Applications', Techniques in Metals Research, Ed. R.F. Bunsah, vol. 2C, Wiley, New York, (1970).

D.C. Joy, J. Micros., **103** (1975) 1.

D. Juul Jensen, Proc. 'Microscale Textures of Materials', Tex. and Mic., in press, (1992).

D. Juul Jensen and V. Randle, Proc. 10th Riso Int. Symp., 'Materials Architecture', Ed. J.B. Bilde-Sorensen et al, Riso Press, Denmark, (1989) 103.

D. Juul Jensen and N.H. Schmidt, Proc. ICOTOM9, Tex. and Mic., **14-18**, (1991) 97.

W.A. Kaysser, S. Takajo and G. Petzow, Z. Metall., **73** (1982) 579.

U. Klement, P. Haasen and P.-J. Wilbrandt, Proc. ICOTOM9, Tex. and Mic., **14-18** (1991) 623.

H. Kokawa, T. Watanabe and S. Karashima, Phil. Mag., **A44** (1981) 1239.

H. Kokawa, T. Watanabe and S. Karashima, Scripta Met., **21** (1987) 839.

C.V. Kopezky and L.K. Fionova, Proc. 'Recrystallisation '90', Ed. T. Chandra, Min. Met. Mat. Soc., USA, (1990) 255.

C.V. Kopezky, A.V. Andreeva and G.D. Sukhomlin, Acta Met. Mat., **39** (1991) 1603.

N.C. Krieger Lassen, D. Juul Jensen and K. Conradsen, Scanning Microscopy, in press (1992).

M.L. Kronberg and F.H. Wilson, Met. Trans., **185** (1949) 501.

K.J. Kurzydlowski, Scripta Met., **24** (1990) 879.

J. Kwiecinski and J.W. Wyrzykowski, Acta Met. Mat., **39** (1991) 1953.

F.F. Lange, Acta Met., **15** (1967) 311.

S. Lartigue and L. Priester, Acta Met. Mat., **11** (1983) 1809.

M.S. Laws and P.J. Goodhew, Acta Met. Mat., **39** (1525).

L.C. Lim and R. Raj, Acta Met., **8** (1984a) 1177.

L.C. Lim and R.Raj, Acta Met., **8** (1984b) 1183.

L.C. Lim and T. Watanabe, Acta Met. Mat., **38** (1990) 2507.

W. Liu, M. Bayerlein, H. Mughrabi, A. Day and P.N. Quested, Acta Met. Mat., **40** (1992) 1763.

W. Lorenz and H. Hougardy, Tex. Mic., **8/9** (1988) 579.

M.H. Loretto, 'Electron Beam Analysis of Materials', Chapman and Hall, London (1984).

J.K. Mackenzie, Acta Met., **12** (1964) 223.

R.A.D. Mackenzie, M.D. Vaudin and S.L. Sass, Proc. Materials Research Society Symp. No. 122, 'Interfacial Properties and Design', Ed. M.H. Yoo et al, MRS, Pittsburg, USA, (1988) 461.

H. Makita, S. Hanada and O. Izumi, Acta Met. Mat., **36** (1988) 403.

H. Makita, S. Hanada, O. Izumi, H. Fukuda and T. Imaizumi, Proc. 'Recrystallisation '90', Ed. T. Chandra, Min. Met. Mat. Soc., USA, (1990) 617.

A. Marrouche, F. Barbier and M. Dechamps, Proc. Conf. 'Ceramic Microstructures '86', Materials Science Research Vol. 21, Ed. J.A. Pask and A.G. Evans, Plenum Press (1987) 231.

H.O. Martikainen and V.K. Lindroos, Acta Met., **33** (1985) 1223.

D.C. Martin and E.L. Thomas, Phil. Mag., **A64** (1991) 903.

J.L. Maurice, J.Y. Laval and J.E. Bouree, J. de Phys., **46** (1985) C4-405.

D. Mckie and C. Mckie, 'Crystalline Solids', Nelson, London, (1974).

M. McLean, J. Mat. Sci., **8** (1973) 571.

D. McLean, 'Grain Boundaries in Metals', Clarendon Press, Oxford, UK, (1957).

K.L. Merkle, Proc. 46th Meeting of the Electron Microscopy Society of America, (1988) Ed. G.W. Bailey, San Francisco Press Inc., 588.

K.L. Merkle, Scripta Met., **23** (1989) 1487.

K.L. Merkle and D.J. Smith, Phys. Rev. Let., **59** (1987) 2887.

K.L. Merkle and D. Wolf, Phil. Mag., **65A** (1992) 513.

A. Morawiec and J. Pospiech, Tex. and Mic., **10** (1989) 211.

N.F. Mott, Proc. Phys. Soc., **A63** (1950) 616.

L.E. Murr, A. Advani, S.Shankar and D.G. Atteridge, Mat. Char., **24** (1990) 135.H. Mykura, in 'Grain Boundary Structure and Kinetics', American Society for Metals, Ohio, USA, (1980), 445.

I. Nakamichi, J. Sci. Hiroshima University, **54A(1)** (1990) 49.

P. Neumann, Proc. ICOTOM9, Tex. and Mic., **14-18** (1991) 53.

C.S. Nichols, R.F. Cook, D.R. Clarke and D.A. Smith, Acta Met. Mat., **39** (1991a,b) 1657 and 1667.

T. Ogura, T. Watanabe, S. Karashima and T. Masumoto, Acta Met., **35** (1987) 1807.

R. Omar and H. Mykura, Proc. Materials Research Society Symp. No. 122, Ed. M.H. Yoo et al, MRS, Pittsburg, USA (1988)61.

R. Orsund, J.Hjelen and E. Nes, Scripta Met., **23** (1989) 1193.

S.R. Ortner and V. Randle, Scripta Met., **23** (1989) 1903.

V. Paidar, Acta Met., **35** (1987) 2035.

G. Palumbo and K.T. Aust, Mat. Sci. Eng., **A113** (1989) 139.

G. Palumbo and K.T. Aust, Acta Met. Mat., **38** (1990a) 2343.

G. Palumbo and K.T. Aust, Scripta Met. Mat., **24** (1990b) 1771.
G. Palumbo and K.T. Aust, Proc. 'Recrystallisation '90', Ed. T. Chandra, (1990c) Min. Met. Mat. Soc., USA, 101.
G. Palumbo and K.T. Aust, 'Materials Interfaces: Atomic Level Structure and Properties', Ed. D. Wolf and S.Yip, Chapman and Hall, (1992) in press.
G. Palumbo, S.J. Thorpe and K.T. Aust, Scipta Met. Mat., **24** (1990) 1347.
G. Palumbo, P.J. King, K.T. Aust, U. Erb and P.C. Lichtenberger, Scripta Met. Mat., **25** (1991) 1775.
G. Palumbo, P.J. King, P.C. Lichtenberger, K.T. Aust and U. Erb, 'Structure and Properties of Interfaces in Materials', Ed. W.A.T. Clark et al, Materials Research Society, Pittsburgh, P.A. (1992) 311.
R. Penelle, T. Baudin, P. Paillard and L. Mora, ICOTOM9, Tex. and Mic., **14-18** (1991) 597.
B. Pledge, in 'Theoretical Methods of Texture Analysis', Ed. H.J. Bunge, DGM Oberursel, Germany (1987) 393.
B. Plutka and H.P. Hougardy, Proc. ICOTOM9, Tex. and Mic., **14-18** (1991) 697.
R.C. Pond and W. Bollmann, Phil. Trans. R. Soc. Lond., **A292** (1979) 449.
R.C. Pond and V. Vitek, Proc. Roy. Soc. Lond., **B357** (1977) 453.
R.C. Pond, V. Vitek and D.A. Smith, Acta Cryst., **A35** (1979) 689.
J. Pospiech, K. Sztwiertnia and F. Haessner, Tex. Mic., **6** (1986) 201.
W. Prantl, E. Werner and H.P. Stuwe, Tex. Mic., **8/9** (1988) 483.
L. Priester, Revue. Phys. Appl., **24** (1989) 419.
P.H. Pumphrey, PhD Thesis, Cambridge, UK, (1974).
P.H. Pumphrey, 'Grain Boundary Structure and Properties', Ed. G.A. Chadwick and D.A. Smith, Academic Press, New York, (1976) 139.
P.H. Pumphrey and K.M. Bowkett, Scripta Met., **5** (1971) 365.
P.H. Pumphrey and K.M. Bowkett, Scripta Met., **6** (1972) 31.
X.R. Qian, Y.T. Chou and E.A. Kamenetzky, J. Less Comm. Met., **134** (1987) 179.
L. Rabet, L. Kestens, P. Van Houtte and E. Aernoudt, Proc. Conf. 'Grain Growth in Polycrystalline Materials', Rome, Mat. Sci. For., **94-96** (1992) 611.
B. Ralph, J. de Phys., **36** (1975) C4-71.
B. Ralph, in 'Grain boundary structure and kinetics', American Society for Metals, Ohio, USA (1980) 181.
B. Ralph, University of Wales Review, **4** (1988) 29.
B. Ralph and R.C. Ecob, in 'Microstructural Characterisation', Ed. N. Hessel Andersen et al, Riso Press, Denmark, (1984) 109.
B. Ralph, P.R. Howell and T.F. Page, Phys. Stat. Sol., **55B** (1973) 641.
B. Ralph, R.C. Ecob, A.J. Porter, C.Y. Barlow and N.R. Ecob, Proc. 2nd Riso Int. Symp., 'Deformation of Polycrystals: Mechanisms and Microstructure', Ed. N. Hansen et al, Riso press, Denmark, (1981) 111.

V. Raman, T. Watanabe and T.G. Langdon, Acta Met., **37** (1989) 705.
V. Randle, Scripta Met., **23** (1989) 773.
V. Randle, Proc. Roy. Soc. Lond., **431A** (1990a) 61.
V. Randle, Met. Trans. **21A** (1990b) 2215.
V. Randle, Mat. Sci. Tech., **6** (1990c) 1231.
V. Randle, Acta Met., **39** (1991a) 481.
V. Randle, Proc. 'ICOTOM9', Tex. and Mic., **14/18** (1991b) 745.
V. Randle, Mat. Sci. Tech., **7** (1991c) 985.
V. Randle, Proc. 'Grain Growth in Polycrystalline Materials', Mat. Sci. For., **94-96** (1992a) 233.
V. Randle, Proc. 'Microscale Textures of Materials,Tex. and Mic., (1992b) in press.
V. Randle, 'Microtexture Determination and its Applications', Inst. of Materials, London, (1992c).
V. Randle and A. Brown, Phil. Mag., **A58** (1988) 717.
V. Randle and A. Brown, Phil. Mag., **A59** (1989) 1075.
V. Randle and D.J. Dingley, Scripta Met., **23** (1989) 1565.
V. Randle and D.J. Dingley, Scripta Met., Proc. 'EUROMAT '89', Ed. H.E. Exner and V. Schumacher, DGM Oberursel, Germany, (1990a) 101.
V. Randle and D.J. Dingley, Proc. 'Recrystallisation '90', Ed. T. Chandra, Min. Met. Mat. Soc., USA, (1990b) 175.
V. Randle and D.J. Dingley, work in preparation, (1992).
V. Randle and J. Furley, Proc. 'ICOTOM9', Tex. and Mic., **14/18** (1991) 877.
V. Randle and B. Ralph, Proc. 'EMAG '85', Inst. Phys. Conf. Ser. No. 78, Adam Hilger, UK, (1985) 59.
V. Randle and B. Ralph, J. Mat. Sci., **21** (1986) 3823.
V. Randle and B. Ralph, J. Mat. Sci., **22** (1987a) 2535.
V. Randle and B. Ralph, Proc. 'EMAG '87', Inst. Phys. Conf. Ser. No.90, Adam Hilger, UK, (1987b) 205.
V. Randle and B. Ralph, Proc. Roy. Soc. Lond., **A415** (1988a) 239.
V. Randle and B. Ralph, J. de Phys., **23** (1988b) 501.
V. Randle and B. Ralph, J. Mat. Sci., **23** (1988c) 934.
V. Randle and B. Ralph, Tex. and Mic., **8/9** (1988d) 531.
V. Randle and B. Ralph, Proc. 46th Meeting of EMSA (Electron Microscopy Society of America, Ed. G.W. Bailey, San Francisco Press, USA, (1988e) 63.
V. Randle and B. Ralph, Proc. Materials Research Society Symp. No. 122, Ed. M.H. Yoo et al, MRS, Pittsburg, USA (1988f) 419.
V. Randle and B. Ralph, Proc. EUREM Ed. P.J. Goodhew and H.G. Dickson, Inst. of Physics Publishing, UK, (1988g) 231.
V. Randle, B. Ralph and D. Dingley, Acta Met., **36** (1988) 267.
S. Ranganathan, Acta Cryst., **21** (1966) 197.
W.T. Read and W. Shockley, Phys. Rev., **78** (1950) 275.

R.J. Roe, J. Appl. Phys., **36** (1965) 2024.
W. Rosenhain and J.C.W. Humphrey, J. Iron Steel Inst., **87** (1913) 319.
N. Rouag and R. Penelle, Tex. and Mic., **11**, (1989) 203.
N. Rouag, G. Vigna and R. Penelle, Acta Met. Mat., **38** (1990) 1101.
N.H. Schmidt, J.B. Bilde-Sorensen and D. Juul Jensen, Scanning Micros. **5** (1991) 637.
R.A. Schwarzer, Tex. and Mic., **13** (1990) 15.
R.A. Schwarzer and H. Weiland, Tex. and Mic., **8/9** (1988) 551.
V.D. Scott and G. Love, Mat. Sci. Tech., **3** (1987) 300.
R. Sharko, Thesis, Paris (1983).
R. Shimizu and J. Harase, Acta Met., **37** (1989) 1241.
R. Shimizu, J. Harase and D.J. Dingley, Acta Met. Mat., **38** (1990) 973.
L.S. Shvindlerman and B.B. Straumal, Acta Met., **33** (1985) 1735.
D.A. Smith, Scipta Met., **8** (1974) 1197.A.P. Sutton and V. Vitek, Phil. Trans. Roy. Soc. Lond. **A309** (1983) 1.
A.P. Sutton and R.W. Balluffi, Acta Met., **35** (1987) 2177.
S. Suzuki, K. Abiko and H. Kimura, Scripta Met., **15** (1981) 1139.
W.A. Swiatnicki, W. Lojkowski and M.W. Grabski, Acta Met., **34** (1986) 599.
K. Sztwiertnia and F. Haessner, Proc. ICOTOM9, Tex. and Mic., **14-18** (1991) 641.
C.J. Tweed, B. Ralph and N. Hansen, Acta Met., **32** (1984) 1407.
P. Van Houtte and F. Wagner, in 'Preferred Orientation in Deformed Metals and Rocks: an Introduction to Modern Texture Analysis', Ed. H.R. Wenk, Academic Press Inc., USA, (1985) 233.
R.A. Varin, Phys. Stat. Sol., **51A** (1979) K189.
J.A. Venables and C.J. Harland, Phil Mag. **27** (1973) 1193.
T. Wang, P.R. Morris and B.L. Adams, Met. Trans., **21A** (1990a) 2265.
T. Wang, B.L. Adams and P.R. Morris, Met. Trans., **21A** (1990b) 2223.
D.H. Warrington and M. Boon, Acta Met., **23** (1975) 599.
D.H. Warrington, in 'Grain Boundary Structure and kinetics', American Society for Metals, Ohio, USA, (1980) 1.
T. Watanabe, Phil. Mag., **47A** (1983) 141.
T. Watanabe, Res. Mech., **11** (1984) 47.
T. Watanabe, Scripta Met., **21** (1987) 427.
T. Watanabe, J. de Phys., **49** (1988) C5-507.
T. Watanabe, Proc. 'Recrystallisation '90', Ed. T. Chandra, (1990) Min. Met. Mat. Soc., USA, 405.
T. Watanabe, K.I. Arai, K. Yoshimi and H. Oikawa, Phil. Mag. Lett. **59** (1989) 47.
T. Watanabe, H. Fujii, H. Oikawa and K.I. Arai, Acta Met., **37** (1989) 941.
T. Watanabe, Y. Suzuki, S. Tanii and H. Oikawa, Phil. Mag. Lett., **62** (1990) 9.

H. Weiland, Acta Met. Mat., in press.
H. Weiland, R.A. Schwarzer and H.J. Bunge, Proc. 8th Int. Conf. on textures of materials (ICOTOM8), Ed. J.S. Kallend and G. Gottstein, The Metallurgical Society, USA, (1988) 953.
H. Weiland, J. Liu and E. Nes, Proc. ICOTOM9, Tex. and Mic. **14-18** (1991) 109.
E. Werner and W. Prantl, J. Appl. Cryst, **21** (1988) 311.
D. Wolf, J. de Phys., **46** (1985) C4-197.
D. Wolf, in 'Ceramic Microstructures '86', Ed. J.A. Pask and A.G. Evans, Plenum Publishing Corporation, (1988) 177.
D. Wolf, Phil. Mag. **A62** (1990a) 447.
D. Wolf, Acta Met. Mat., **38** (1990b) 791.
D. Wolf and J.F. Lutsko, Zeit. Kristall., **189** (1989) 239.
S.I. Wright, J. Zhao and B.L. Adams, Tex. and Mic., **13** (1991) 123.
S.I. Wright and B.L. Adams, Proc. ICOTOM9, Tex. and Mic. **14-18** (1991) 273.
S.I. Wright and B.L. Adams, Proc. 'Microscale Textures of Materials', Tex. and Mic., in press (1992).
J.W. Wyrzykowski and M.W. Grabski, Phil. Mag. **53A** (1986) 505.
C.T. Young, J.H. Steele and J.L. Lytton, Met. Trans., **4** (1973) 2081.
J. Zhao and B.L. Adams, Acta Cryst., **A44** (1988) 326.
J. Zhao, B.L. Adams and P.R. Morris, Tex. and Mic., **8/9** (1988) 493.
J. Zhao, J.S. Koontz and B.L. Adams, Met. Trans. **19A** (1988) 1179.
Y. Zhu, H. Zhang, H. Wang and M. Suenaga, J. Mat. Res., **6** (1991) 2507.

INDEX